PROGRESS

Infinite Universe Theory

Glenn Borchardt

Being a suggested replacement for the Big Bang Theory, a call for the demise of cosmogony, and the initiation of the Last Cosmological Revolution.

Published by the Progressive Science Institute

P.O. Box 5335

Berkeley, CA 94705

USA

Cover: A portion of our galaxy, the Milky Way, which is estimated to contain about 400 billion stars similar to our Sun. With over two trillion galaxies so far observed, this is less than 0.00000000005% of the known universe. Credit: NASA.

Copyright © 2017 by Glenn Borchardt

ASIN: B076X85XVN (ebk)

ISBN: 9781973399056 (pbk)(b&w)

ISBN: 9781729092521 (pbk)(color)

Citation: Borchardt, Glenn, 2017, Infinite universe theory (Version 20190331): Berkeley, CA, Progressive Science Institute, 349 p.

Also by Glenn Borchardt

The Ten Assumptions of Science

The Scientific Worldview

Universal Cycle Theory (with Stephen J. Puetz)

Acknowledgements

I thank Marilyn Borchardt, Roger Burbach, Fred Frees, Bill K. Howell, and numerous Blog commenters for many stimulating discussions of the topic. Of course, I am especially indebted to Stephen J. Puetz, my coauthor on "Universal Cycle Theory," the preceding technical volume. Without his outstanding and sagacious input, I doubt we ever would have discovered the physical cause of gravitation. Thanks so much to the following reviewers who provided suggestions that improved the manuscript: Marilyn Borchardt, Rick Dutkiewicz, Steven B. Bryant, Luis Cayetano, Gerald Ardu, William Westmiller, George Conger, Duncan Shaw, Stephen Puetz, Bill Howell, James Wright, Mike Gimbel, Ed Mason, Jesse Witwer, Juan Calsiano, Curt Weinstein, and Piotr Lapinski.

During the last half century, the following institutions generously provided support for my scientific career: University of Wisconsin, Wisconsin Geological & Natural History Survey, Oregon State University, United States Atomic Energy Commission, National Science Foundation, National Academy of Sciences, National Research Council, United States Geological Survey, California Geological Survey, United Nations, and numerous private clients. I wish to thank those who spent endless hours administering these organizations so I would be free to enjoy my explorations in the laboratory and field. I also thank the taxpayers and clients who provided the funds that made these investigations possible. In gratitude, I present this book at minimal cost.

I dedicate this book to:

Marjorie, Arnold, Bertha, Coutchie, Roger, Marilyn, Jim, Marion, Francis, Rod, Moyle, Art, Elizabeth, Dennis, Elia, Nina, Hasu, Chuck, Doug, Karl, Harry, Edward, Fred, Bob, Tom, Natalie, Steve, Jesse, and Juan.

Infinite Universe Theory
Glenn Borchardt

Contents

Acknowledgements *4*

List of Tables *10*

List of Figures *11*

Preface *18*

Introduction *21*

Part I: What is wrong with the Big Bang Theory? **30**

Chapter 1 .... *31*

Immensity *31*

Chapter 2 .... *34*

Galaxies in Collision *34*

Chapter 3 .... *37*

Galaxy Clusters in Collision *37*

Chapter 4 .... *39*

Elderly Galaxies at the Edge of the Universe *39*

Chapter 5 .... *45*

Elderly Galaxy Clusters at the Edge of the Universe *45*

Chapter 6 .... *49*

Solipsism and Perception *49*

Chapter 7 .... *53*

Einstein's "Untired Light Theory" *53*

Chapter 8 .... *58*

Space-time Salvation 58

Part II: Infinite Universe Theory 61

Chapter 9 65

The Ten Assumptions of Science 65

Chapter 10 85

Progressive Physics 85

Chapter 11 96

Neomechanics 96

Chapter 12 135

Univironmental Analysis 135

Part III: Questions resolved by Infinite Universe Theory 147

Chapter 13 150

Scientific Philosophy 150
   13.1 Does curiosity imply the advent of Infinite Universe Theory? 150
   13.2 Does calculus imply the advent of Infinite Universe Theory? 154
   13.3 Does univironmental determinism imply the advent of Infinite Universe Theory? 155
   13.4 What is a BS meter and why do you need one? 156
   13.5 Does Infinite Universe Theory resolve the "Who Created God" question? 156

Chapter 14 158

Regressive Misconceptions 158
   14.1 Did MMX prove that aether did not exist? 158
   14.2 Is the speed of light constant? 159
   14.3 Does energy have mass? 164
   14.4 Does "dark energy" exist? 169
   14.5 Does the "god particle" exist? 171
   14.6 Is string theory valid? 175
   14.7 Does space-time exist? 176
   14.8 Is matter a result of quantum fluctuations? 182

14.9 Does the double-slit experiment prove light is both a wave and a particle? ...184
14.10 Does Infinite Universe Theory mean everything is possible? ...187
14.11 Is the "Twin Paradox" Resolvable without Relativity? ...188

Chapter 15 ...*191*

Tests of Relativity ...*191*
15.1 Did Sagnac prove the existence of aether? (1913) ...193
15.2 Did de Sitter prove aether exists? (1913) ...202
15.3 Did Eddington prove space was curved? (1919) ...204
15.4 Did Eddington prove light is affected by gravitation? (1919) ...208
15.5 Does the gravitational redshift confirm relativity? (1960) ...209
15.6 Did clocks flown around Earth confirm relativity? (1972)...217
15.7 Did LIGO prove there are gravitational waves? (2016) ...227

Chapter 16 ...*231*

Progressive Physics ...*231*
16.1 Why can there be no matter without motion? ...231
16.2 What is aether? ...234
16.3 What causes gravitation? ...239
16.4 Where does matter come from? ...249
16.5 What is the cause of charge? ...256
16.6 What is the cause of magnetism? ...258
16.7 Why do satellites stay in orbit? ...260
16.8 Why is there so much spookiness in quantum mechanics? ..263
16.9 Why does matter prevent the transmission of aether waves? ...265
16.10 Why would a finite particle be impossible? ...265
16.11 Does dark matter exist? ...266
16.12 Is matter lost during atomic fission? ...268

Chapter 17 ...*269*

Cosmology ...*269*
17.1 Is the universe expanding? ...269
17.2 Can the Doppler Effect occur without a medium? ...275
17.3 What causes the cosmological redshift? ...275
17.4 Are there galaxies more than 13.8 billion years old? ...278
17.5 What causes the Shapiro Effect? ...282
17.6 Will the universe suffer "heat death"? ...283

17.7 Why is a finite universe impossible?..285

Part IV: Conclusions **287**

Chapter 18 ....*288*

Predictions, Persistence, and Requiem..*288*
    18.1 Predictions of the New Theory..288
    18.2 Paradigmatic Persistence of the Big Bang Theory 289
    18.3 Requiem for the Big Bang Theory 295
    References..299
    Glossary 311
    Appendix..345

List of Tables

Table 1 Proponents of the Big Bang Theory: From priest to Pope. 22
Table 2 Opponents of the Big Bang Theory: From plasma to Universal Cycle Theory. 23
Table 3 The Ten Assumptions of Science. 24
Table 4 Falsifications, contradictions, and paradoxes disproving the Big Bang Theory. 60
Table 5 Characteristics of classical mechanics, neomechanics, and relativity. 134
Table 6 Einstein's eight ad hocs. 160
Table 7 Unpublished original test results and the Hafele and Keating alterations obtained by Kelly in 2000 (nanoseconds). 221
Table 8 Gravitational and kinematic contributions to the H-K results. 221
Table 9 Phenomena considered the result of "gravitational time dilation" (stationary clocks at high altitude). 224
Table 10 Phenomena considered the result of "kinematic time dilation" (moving clocks). 225
Table 11 Summary of speculative calculations on the properties of aether particles and the aether medium (see Appendix for details about how these were calculated from Planck's constant and the known electron mass). 236

List of Figures

Figure 1 A fraction of the 4,500 stars one might see with the naked eye from any one place on Earth. The white streak is a portion of the Milky Way. Credit: Stellarium.. 31

Figure 2 Image showing some of the 50,000 galaxies in the nearby universe detected by the Two Micron All Sky Survey (2MASS) in infrared light. Note that astronomers now estimate there are over two trillion galaxies, so this is less than 0.000003% of them. Credit: T.J. Jarrett (IPAC/Caltech). 33

Figure 3 The Andromeda Galaxy. Credit: NASA. 34

Figure 4 Spiral galaxies NGC 2207 and IC 2163 in collision. .. 35

Figure 5 An explosion has only divergence, while creation requires convergence. Credit: Vector Graphics...... 36

Figure 6 Colliding galaxy clusters. Credit: Chandra X-ray Observatory. 38

Figure 7 NASA's official view of what the Big Bang universe should look like (seriously). Credit: NASA. 42

Figure 8 There are over 10,000 galaxies seen in this photo of the Hubble Ultra Deep Field. It comprises a tiny patch on the sky 1/10 the size of the Moon. Credit: NASA, ESA, H. Teplitz and M. Rafelski (IPAC/Caltech), A. Koekemoer (STScI), R. Windhorst (Arizona State University), and Z. Levay (STScI). 43

Figure 9 Close-up of a small portion of the HUDF. Note that these objects are various colors. Most are not red as implied by the misnomer "cosmological redshift." Color is determined by frequency, not wavelength. 44

Figure 10 "This image shows the Milky Way and Andromeda galaxies in space to scale. It illustrates both the size of each galaxy and the distance between the two galaxies..." ly = light year = 9 trillion km = 5.58 trillion miles. Credit: Jan van der Crabben............ 48

Figure 11 Typical redshift vs. distance plots calculated as erroneously assumed recessional velocities. This is part of an animation prepared by the Institute for Astrophysics and Space Science, Western Kentucky University. 52

Figure 12 Ripple wavelength increases with distance. Credit: Exo.net. 55

Figure 13 ACCELERATION OF THE MICROCOSM. A relatively high velocity supermicrocosm (external microcosm) collides with and transfers motion to a relatively low velocity microcosm. As a result, the microcosm moves to the right with increased velocity. Dashed lines indicate former locations of the impacting supermicrocosm ("super" because it is outside the microcosm, not because it is anything special). 116

Figure 14 Video showing acceleration and deceleration demonstrated by pendulums at http://go.glennborchardt.com/pendulums. Credit: httprover. 116

Figure 15 DECELERATION OF THE MICROCOSM. The microcosm collides with and transfers motion to a low velocity supermicrocosm. As a result, the microcosm tends to move to the left losing part of its rightward velocity while the supermicrocosm moves to the right. 118

Figure 16 ABSORPTION OF MOTION. A high-velocity supermicrocosm collides with and transfers motion to a low-velocity submicrocosm (internal microcosm). As a result, both the microcosm and the submicrocosms inside it are accelerated slightly to the right. 121

Figure 17 The collision between a hammer and a nail cause both to become warm. Credit: Mark Wasteney at flickr.com............ 123

Figure 18 EMISSION OF MOTION. A high-velocity submicrocosm collides with and transfers motion to a low-velocity supermicrocosm. As a result, both the microcosm and the submicrocosms inside it are decelerated slightly to the left............ 124

Figure 19 ABSORPTION OF MATTER. A supermicrocosm enters a low velocity microcosm and becomes a submicrocosm. The supermicrocosm also could be absorbed on the surface of the microcosm, thus widening the microcosmic boundary............ 127

Figure 20 EMISSION OF MATTER. A submicrocosm leaves a low velocity microcosm and becomes a supermicrocosm............ 128

Figure 21 Demonstration of the Second Law of Thermodynamics, Newton's First Law of Motion, and the tendency for microcosms to move toward univironmental equilibrium. Opening the valve will allow the gas molecules in A to move into vacuum chamber B under their own inertia............ 129

Figure 22 Pope disses curiosity. Credit: Slide courtesy of Jerry Coyne and Time............ 151

Figure 23 Hey all you god-fearing teenagers: Don't reason! Just obey! Credit: Slide courtesy of Jerry Coyne. 152

Figure 24 Neomechanical interactions apropos the $E=mc2$ equation illustrating that both absorption and emission involve mechanical collisions described by Newton's Second Law of Motion (from Figure 16 and Figure 18). By denying aether exists, regressive physics denies that these collisions occur............ 166

Figure 25 Infrared photo of radiation from chimps. Credit: Cool Cosmos/IPAC, NASA/JPL-Caltech............ 167

Figure 26 Higgs-Boson propaganda proffered by Time. 173
Figure 27 Cosmic microwave background radiation measured by the Wilkinson Microwave Anisotropy Probe (WMAP) showing the heterogeneous/homogeneous nature of intergalactic temperature. The light areas are greater than 2.7°K and the dark areas are less than 2.7°K, although the variation is tiny: 5 X 10^{-5} °K. 178
Figure 28 Wave behavior of light demonstrated by the double slit experiment (Credit: Steven B. Bryant). 185
Figure 29 Expected result if light was a particle (Credit: Steven B. Bryant). 185
Figure 30 "How interference works. The gap between the surfaces and the wavelength of the light waves are greatly exaggerated. The distance between the dark fringe (a) and the bright fringe (b) indicates a change in the gap's thickness of 1/2 the wavelength." Credit: Chetvorno. Note that the path length on the bottom left is a half cycle longer than the path length on the bottom right. If the two lengths were identical, there would be no fringe. 194
Figure 31 "When the interferometer starts to rotate clockwise the clockwise propagating laser beam has to cover more distance because the detector is moving away. The counter-clockwise propagating beam has to cover less distance as the detector is moving towards it. Now the two laser beams have a different phase at the point where they interfere resulting in a different amplitude of the signal at the detector." Credit: Berhard Albrecht. 195
Figure 32 Olympic pool with movable end. 198
Figure 33 Sid Harris's classic cartoon illustrating scientific skepticism toward an ad hoc. Credit: ©Sidney Harris (sciencecartoonsplus.com). 200

Figure 34 Orbits of binary stars A and B and the velocities expected if light was a particle. 203
Figure 35 Light waves from distant stars bend only in the plasma rim of the Sun due to refraction. They are unaffected by gravitation, contrary to the predictions of relativity (from Dowdye, 2012). 207
Figure 36 This plate from the Eddington paper is a half-tone reproduction from one of the negatives taken with a 4" lens at Sobral, Brazil. The corona prevented any observation of light bending in the plasma rim at the surface of the Sun (from Dyson, Eddington, and Davidson, 1920). 208
Figure 37 How signal transmission through a medium affects ***measurements*** of objects in motion. The path from point A to point B is simply lengthened when B is in motion, producing a measured increase in distance and the time required for the transmission to occur). Credit: Sacamol. 223
Figure 38 Kinematic time difference experiment. The frequency (f_o) decreases as a function of velocity (v_{rms}) when the clock is in motion.. 226
Figure 39 Experimental setup for LIGO. Credit: LIGO CalTech. 228
Figure 40 The Sombrero Galaxy (M104). Does an aether particle look like this vortex disc? Credit: HST/NASA/ESA. 237
Figure 41 Interferometer measurements of Earth's velocity around the Sun as determined at various altitudes above mean sea level. The three data points at high altitude are projections and are yet to be performed. The other data are from Galaev, who seems to be the first to show this relationship............................. 239
Figure 42 Aether particle losing velocity upon colliding with baryonic matter. 243

16   *Infinite Universe Theory*

Figure 43 Newton's push theory of gravitation in which he hypothesized a universal medium with increasing distal density. 246
Figure 44 Earth's atmosphere. 248
Figure 45 Earth-Moon gravitational discs. 249
Figure 46 Microcosms in motion. Note that large microcosm A in the center shelters microcosm B from impacts from the left. Consequently, B will be pushed toward A, with the likelihood it might even end up rotating around A or combining with it. 251
Figure 47 Balloons pushed to the side of the pool illustrating the tendency for microcosms to be pushed together by impacts from the macrocosm. 252
Figure 48 Hypothetical aether particles showing the effects of vortex morphology. The two parallel vortices, each having exposure on one side, will receive fewer impacts than the others will. Credit: Sombrero galaxy images modified from NASA. 254
Figure 49 Demonstration of univironmental "attraction" and repulsion by Ionel Dinu at http://go.glennborchardt.com/Dinucharge. 256
Figure 50 Diagram of how spin direction produces "attraction" and repulsion. 257
Figure 51 Geosynchronous satellites at about 36,000 km above Earth. Credit: Lookang. 261
Figure 52 When the source is in motion to the left, wavelengths are blueshifted (shortened) to the left and redshifted (lengthened) to the right. 273
Figure 53 Cosmological redshift showing spectral lines for various elements being shifted to the red, long wavelength, low energy end of the spectrum. Credit: Georg Wiora. 276
Figure 54 Cover of our book showing the observable universe rotating around the "Local Mega-Vortex." We based

on observations that galaxy clusters flow in a preferred direction, as if they were revolving around some sort of massive celestial body outside the observable universe. 281

Figure 55 Google Ngram for books that use the phrases or words: expanding universe, big bang theory, or multiverse. Note that the popularity of the expanding universe interpretation preceded the Big Bang Theory by over three decades. 293

Figure 56 Sigmoidal growth curve for global population assuming perfect symmetry about the 1989 Inflection Point. 296

Figure 57 Ngram for the words "cosmogony" and "cosmogeny". 314

Preface

When I first became aware of the Big Bang Theory of the universe, I thought "Wow! That's great; finally they know how it all began!" Mom, a devout and very conservative Missouri Synod Lutheran, opposed the whole idea of it—her world began with Genesis 6,000 years ago. At billions of years, the timing of the Big Bang was a bit off for her. With an eighth-grade education, she would not be expected to appreciate how fast light travels and how far away those stars and galaxies were. On the other hand, I graduated from a university that favored "fearless sifting and winnowing by which alone the truth can be found." I had even gotten my nose out of the books long enough to look around me. After learning a bit of soil science, I returned home to southern Wisconsin to see the evidence first hand for an Earth much older than 6,000 years. According to geologists at the university, it turns out I grew up next to a drumlin, an egg-shaped hill that once was under a mile of ice. Each one, so they said, was over 15,000 years old.

Then I found it. Someone carbon-dated peat from one of the many marshes between the drumlins. The date was about 15,000 years. As a literalist born of the Lutheran tradition, I had to make a choice: science or Genesis. I subsequently gathered with my own hands hundreds of samples of peat, wood, and charcoal, with many turning out to be much older than that. Some of the samples dated by other means were millions of years old. During my Postdoc, I even had some colleagues down the hall who dated samples from the Moon at 4.66 billion years old. Guess I chose correctly.

I bring this up because it is similar to the journey I would like you to take in this book. You should begin with a lot of skepticism. After all, what could a farm kid from Wisconsin who plays around in the dirt say about the current theory of the universe? Those smart fellows like Einstein and Hawking surely must know what they are doing. That is what I thought too—up until 1978.

But that was not to be the case after I started looking into their wild claims in depth. How could the universe explode out of nothing? In my scientific experience, everything came from some other thing. Nothing just popped up out of nowhere. How could the universe be 4-dimensional? Everything I knew had only 3-dimensions. How could the universe be expanding in all directions at once? The rubber sheets and claims of curved empty space did not appeal to me. How could light be both a particle and a wave at the same time? In Physics 1a, I was told to leave common sense behind. You needed to know some exceedingly advanced higher math, and besides, only a few physicists could comprehend it anyway. It looked like the emperor's new clothes.

Being skeptical and commonsensical by nature, I investigated "modern physics" systematically from top to bottom. What I found was shocking. The whole thing was an enormous, stinking can of dead worms. The most dubious proclamations were founded on presuppositions that had more in common with religion than science. No wonder modern physics was so popular. I would need to uncover the assumptions that were making physics go awry for over a century. From there I would need to rebuild much of its structure, starting with classical mechanics and incorporating what was missing: the assumption of *infinity*. Happily, not much of this involves complicated math, but if you want to understand this new view of the universe, you need to do some work. In particular, ingrained presuppositions are hard to change. Still, I promise it will be worth it.

Berkeley, December 21, 2017   Glenn Borchardt

The time will come when all will see what I see.-Bruno, 1600[1]

Introduction

Everything we know has a beginning. For millennia, humans have applied that observation to everything that exists—the universe. In most societies, thinkers create myths great and small to answer the question: "Where did it all come from?" Cosmologists, of course, study the universe. I found out the ones who assume the universe had a beginning actually are *cosmogonists*.[2] It is an especially unpopular term, but one I will use throughout this book because that is precisely what they are, even though they are reluctant to admit to that prejudice. The beginning envisioned by today's cosmogonists is a special kind of beginning. Any creation we are used to, a house for instance is the result of convergence, the bringing together of the necessary materials—not so with the modern cosmogonist's universe. There is no convergence for, true to cosmogony, there is nothing to bring together. Today's universe is simply a creation of something out of nothing. The currently popular mythology, the Big Bang Theory, goes a step further. Instead of a convergence, it speaks of a grand divergence: the oxymoronic belief the universe could come together by coming apart.

As fantastic as this explosion of the entire universe out of nothing may seem, it is still considered a real possibility, especially among supporters of traditional beliefs and modern physics (Table 1). However, unlike other theories, such as those involving evolution and plate tectonics, relativity and the Big Bang Theory continue to have detractors within the scientific community itself (Table 2). That is why "proofs" of relativity

---

[1] Bruno, 1600, Giordano Bruno quotes [http://go.glennborchardt.com/Brun]. Said before the Catholic Church burned him at the stake for proclaiming that the universe was infinite.
[2] Words defined in the glossary are hyperlinked. Offsite URLs were accurate as of 20171028. The rebrands (e.g., go.glennborchardt.com) add an additional level of security to those links.

and the Big Bang Theory get attention in the press. The better-established theories demonstrate their validity with each new breed of dog and with each new subduction earthquake. Not so with the Big Bang Theory. No matter how you cut it, the explosion of the universe out of nothing is as unprovable as it is mind-boggling.

Table 1 Proponents of the Big Bang Theory: From priest to Pope.

Author	Date	Title
Lemaître, G.	1950	The primeval atom: An essay on cosmogony
Gamow, George	1954	Modern cosmology
Gamow, George	1961	The creation of the universe
Bergamini, D.	1962	The universe
Hawkins, G.S.	1962	Expansion of the universe
Silk, Joseph	1973	Cosmological theory
Gott, J.R., III	1976	Will the universe expand forever?
Wald, R.M.	1977	Space, time, and gravity: The theory of the big bang and black holes
Jastrow, Robert	1978	God and the astronomers
Silk, Joseph	1980	The big bang: The creation and evolution of the universe
Davies, P.C.W.	1981	The edge of infinity: Where the universe came from and how it will end
Hawking, S.W.	1988	A brief history of time
Silk, Joseph	2002	The big bang (3rd ed.)
Hawking, S., & Mlodinow, L.	2012	The grand design
Krauss, L.M.	2013	A universe from nothing
McKenna, J.	2014	Pope says evolution, Big Bang are real

The title of this book promises a more rational approach, though from the current perspective, it too is mind-boggling. It is based on the opposite view I called number Eight in my Ten Assumptions of Science: ***infinity***, which assumes the universe is infinite, both in the microcosmic and the macrocosmic directions

(Table 3). We face a choice between two opposing, unprovable assumptions: *finity*, which implies a beginning, and ***infinity***, which does not. Stephen Puetz and I used ***infinity*** and the other assumptions to write an extensive tome developing the detail supporting the case for a hierarchically infinite universe.[3] It is

Table 2 Opponents of the Big Bang Theory: From plasma to Universal Cycle Theory.

Author	Date	Title
Alfvén, Hannes	1977	Cosmology, History, and Theology
Narliker, Jayant	1981	Was there a big bang?
Borchardt, Glenn	1984	The scientific worldview
Marmet, Paul	1990	Big Bang Cosmology Meets an Astronomical Death
Lerner, E.J.	1992	The Big Bang never happened
Mitchel, W.C.	1994	The cult of the big bang: Was there a bang?
Farmer, B.L.	1997	Universe alternatives
Arp, Halton	1998	Seeing red
Martin Jr., R.C.	1999	Astronomy on Trial
Mitchel, W.C.	2002	Bye bye big bang: Hello reality
Borchardt, Glenn	2004	Ten assumptions of science and the demise of 'cosmogony'
Ashmore, Lyndon	2006	Big Bang Blasted
Borchardt, Glenn	2007	The Scientific Worldview: Beyond Newton and Einstein
Disney, M.J.	2011	Doubts about Big Bang Cosmology
Puetz, S.J. and Borchardt, Glenn	2011	Universal cycle theory: Neomechanics of the hierarchically infinite universe

---

[3] Puetz and Borchardt, 2011, Universal Cycle Theory. An application of Infinite Universe Theory to vortices and cycles from the infinitely small to the infinitely large.

Table 3 The Ten Assumptions of Science.[4]

1.	Materialism	The external world exists after the observer does not.
2.	Causality	All effects have an infinite number of material causes.
3.	Uncertainty	It is impossible to know everything about anything, but it is possible to know more about anything.
4.	Inseparability	Just as there is no **motion** without **matter**, so there is no matter without motion.
5.	Conservation	Matter and the motion of matter can be neither created nor destroyed.
6.	Complementarity	All things are subject to divergence and convergence from other things.
7.	Irreversibility	All processes are irreversible.
8	Infinity	The universe is infinite, both in the microcosmic and macrocosmic directions.
9.	Relativism	All things have characteristics that make them similar to all other things as well as characteristics that make them dissimilar to all other things.
10.	Interconnection	All things are interconnected, that is, between any two objects exist other objects that transmit matter and motion.

over 600 pages, with over 40 pages of references alone. You will want to read it if you are scientifically and mathematically inclined. It has a good review of cosmology from the univironmental perspective. In particular, we are most proud of our early work suggesting that aether was involved in the physical cause of gravitation,[5] which became a logical consequence of its compilation and the modifications I propose below as Aether Deceleration Theory.

---

[4] Borchardt, 2004, The ten assumptions of science (These will be summarized and discussed in a later chapter.)
[5] Borchardt and Puetz, 2012, Neomechanical Gravitation Theory.

Although not generally popular with the church, the idea the universe is infinite has been around for over five centuries (*a la* Bruno). Three centuries ago, Isaac Newton mentioned the universe must be infinite or else everything would be attracted to one spot by gravitation.[6] Unfortunately, most of today's objectors to the Big Bang Theory fail to support Infinite Universe Theory as its obvious replacement. They may detest the idea of the universe exploding out of nothing, but cannot take that final, antireligious step toward the idea the universe exists everywhere and for all time without any supernatural help. Many are reluctant to challenge the ever-popular theory of relativity, which is the foundation of the Big Bang Theory. Others propose oxymoronic multiverses, each presumably having its own explosion from nothing. Attempts at reform are doomed to failure because the Big Bang Theory is illogical at its core. At the Progressive Science Institute (PSI), Infinite Universe Theory has been hatching since 1978 when I gave up on the Big Bang Theory, switching assumptions from *finity* to **infinity**. At first, I tried to work with relativity, believing like many reformers, that photons existed and aether did not. Subsequent books and papers on scientific philosophy, physics, and the vagaries of cosmogony produced much progress. The present book attempts to reach a wide audience. As such, I will avoid much of the mathematics only a scientist could love. Note in Part I, I will only cover a few of the most obvious contradictions plaguing the present cosmogony. Numerous authors already have voiced strong objections to Big Bang Theory (Table 2). The table shows the Big Bang Theory has been "falsified" (proven false) on numerous occasions, notably without the publicity anathema to a

---

[6] Newton, 1688, Letter to Dr. Covel. [Note that Newton was not consistent on macrocosmic infinity. He assumed *finity* in a letter written to Richard Bentley in 1692, he speculates that the universe will "not end before 2060. .... It may end later, but I see no reason for its ending sooner."]

reigning paradigm. Nevertheless, as Kuhn, the famous historian of science, observed, mere falsification is not enough to unseat a powerful paradigm.[7] There are always long-time "true believers" who are enamored with a particular idea and will not change despite the evidence.[8] Even the Flat Earth Theory is suitable for a drive to Los Angeles. But finite theories inherently contain a ticking time bomb: only applying to finite portions of the universe. Eventually finite theories overreach into territory to which they are not suited. The Flat Earth Theory did not get us to the Moon.

An alternate theory must resolve the numerous paradoxes and contradictions of the Big Bang Theory, while producing successful predictions. That is the purpose of the Puetz-Borchardt treatise. Unfortunately, the Big Bang Theory has great popular support, with a religiously flavored philosophical foundation generally opposed to the Ten Assumptions of Science. Coincidentally, that indeterministic foundation provided an opening for Georges Lemaître, a Jesuit priest who suggested the expansion of the universe meant it exploded from a "cosmic egg" in tune with Genesis.[9] The idea was taken so seriously by its indeterministic supporters that an opponent, Sir Fred Hoyle, facetiously coined the term "Big Bang" during a BBC radio broadcast on 28 March 1949.[10] As far-fetched as it seems, the Big Bang Theory amounts to the last big compromise between science and religion. As such, I do not expect the interconnection between them to be broken easily. It will take decades before the Big Bang Theory and its foundational theories succumb. In the

---

[7] Kuhn, 1962, 1996, [2012], The Structure of Scientific Revolutions.
[8] I observed this personally when plate tectonic theory became well accepted after 1966. Several older geologists who were leaders of the California Geological Survey or prominent professors at U.C. Berkeley took their disbelief to their graves.
[9] Lemaître, 1950, The primeval atom: An essay on cosmogony.
[10] Mitton, 2011, Fred Hoyle: A life in science.

meantime, there is no reason for you to delay putting yourself at the forefront of a debate that will be won by Infinite Universe Theory, the only possible alternative to the Big Bang Theory.

By the nature of the Infinite Universe, this alternative approach must present one huge circular argument. That is mere logic: begin with infinity and end with infinity as I do, or begin with finity and end with finity as the cosmogonists do. Because the universe is infinite, both infinity and finity must forever remain assumptions. Assumptions, if they are fundamental, must have opposites and must be unprovable.[11] No one is ever going to travel to the "end of the universe" to see whether it is infinite or not. The best we can do is to assume one or the other. This book jumps on the merry-go-round of *infinity* to see where it takes us. What kind of universe would it be if it were infinite? Of course, if you are more conventional and averse to taking that ride, then this book is not for you. Its few pages are not going to convince true believers, such as the physicists and cosmogonists who built their finances around *finity*. On the other hand, if you have an open mind, I think you will appreciate the arguments derived from deductive reasoning. It makes a nice tidy package, sewing up many loose ends left untied by *finity*-related theories. It resolves the Big Bang Theory's paradoxes and contradictions, answers questions untouchable by cosmogony, makes predictions, and suggests experiments never offered by the Big Bang Theory.

Before we can produce our little "out with the old, in with the new" stunt, we need to review the Big Bang Theory. What better way to do this than to quote the conventional wisdom promulgated by those who take the time to write for Wikipedia? Note the seriousness, certitude, and overall hubris with which the authors present this hypothesized "explosion of everything from

---

[11] Collingwood, 1940, An essay on metaphysics.

nothing." As with any theory worth its salt, promoters interpret the observations and experiments mentioned from this solitary point of view. After much repetition, what is mere speculation based on dubious foundational assumptions becomes "objective" truth. They dare not mention contrary data and falsifications. These fall by the wayside in the same way we tend to forget the "cons" as soon as we make a difficult and important decision. This is not conspiratorial; it is only human—to make a decision, we need closure. Our choices of a spouse or automobile best not dwell on the negatives after the fact. It is not efficient to wake each morning searching for a new spouse and a new automobile. Also, remember to make a decision, we need to have closure. Thus, we must "close our minds" to other possibilities. Closure reduces cognitive dissonance and makes our lives simpler. That is why the current batch of physicists and cosmologists will detest this book. Like Newton's object once in motion, we favor least motion, which allows us to go humming down life's track with the least effort. We still need to make millions of decisions, but the ones having already experienced closure need not be among them.

Here is the promised overview of the Big Bang Theory according Wikipedia. Although we are warned Wikipedia articles may be rife with errors, I include this reference anyhow because it appears to represent the views of its promoters at this time. You can check the source articles for errors in interpretation. This is the leading paragraph:

> The **Big Bang** theory is the prevailing cosmological model for the universe from the earliest known periods through its subsequent large-scale evolution. The model describes how the universe expanded from a very high density and high temperature state, and offers a comprehensive explanation for a broad range of phenomena, including the abundance of light elements, the cosmic microwave background, large scale structure and Hubble's Law. If the known laws of

physics are extrapolated to the highest density regime, the result is a singularity which is typically associated with the Big Bang. Physicists are undecided whether this means the universe began from a singularity, or that current knowledge is insufficient to describe the universe at that time. Detailed measurements of the expansion rate of the universe place the Big Bang at around 13.8 billion years ago, which is thus considered the age of the universe. After the initial expansion, the universe cooled sufficiently to allow the formation of subatomic particles, and later simple atoms. Giant clouds of these primordial elements later coalesced through gravity in halos of dark matter, eventually forming the stars and galaxies visible today.[12]

Well, there you have it, the Big Bang Theory in all its glory. Someday, that Wikipedia quote will be of great historical significance. In the meantime, let us proceed with its replacement, beginning with a few glaring contradictions purposely left out of the overview above. Once we switch to Infinite Universe Theory, those observations will no longer be contradictory. They will fit right in with the story I am about to tell. You will find the result to be much more satisfactory. In particular, Infinite Universe Theory has no contradictions and no paradoxes. That is because the assumption of ***infinity*** is radically different from the cosmogonists' assumption of *finity* and the reformists' oxymoronic attempts to merge the two assumptions in their efforts to appease doubters. As you will see, nothing less than a clean break with our indeterministic past can rid us of the gnawing concern there indeed must be something wrong with the idea the entire universe exploded out of nothing. In this book, you will learn why nonexistence is impossible; why the universe must extend in all directions with no stopping point; and why the universe has always existed and will continue to exist forever.

---

[12] http://go.glennborchardt.com/WikiBBT [Accessed on 20171030].

PART I: WHAT IS WRONG WITH THE BIG BANG THEORY?

Chapter 1
Immensity

If the universe is finite, how would we know it? If we only saw a few planets and stars, as in ancient times, we might conclude that easily. After all, even on a good night we cannot see more than 9096 stars with the naked eye.[13] At any one time, one side of Earth blocks out half of those, so the maximum anyone can see on the ground is about 4,500 stars (Figure 1). That still might be a manageable number for those inclined to believe the universe had an origin, and therefore had to be finite in extent. Indeed, for a long time people thought the stars were fixed to a sphere. That explained why they seemed to rotate *en mass* around the North Pole, forming a nice little cocoon made especially for us. Of course, the rotation was due to the Earth, not of any imagined sphere. Then there was the little problem of the whitish streak that always showed up too.

Figure 1 A fraction of the 4,500 stars one might see with the naked eye from any one place on Earth. The white streak is a portion of the Milky Way. Credit: Stellarium.

That streak is typical of what happens when we study the Infinite Universe. Once we think we finally figured it out, something else turns up. The whitish streak was evidence for the

---

[13] King, 2014, 9,096 Stars in the Sky.

"Milky" Way, the much larger conglomeration of which we are a part (cover photo). Once we had telescopes, that white streak obviously contained stars—now estimated to number about 400 billion,[14] many like our Sun, with its own satellites, and some with habitable planets like Earth. That is the way the Infinite Universe works: no matter what we find, there always has been more. The Milky Way is a spiral galaxy rotating around a central core, which although much more populated than expected, was a supposedly self-contained system—until improvements in telescopes proved there were other so-called "island universes"—galaxies similar to ours (Figure 2). The improvements never stopped, and with each major advance, we found more galaxies. At last count, some astronomers estimate there are at least ***two trillion galaxies***[15] and at least ***800 billion trillion stars!***[16] I don't know about you, but that seems awfully close to being infinite right there. The Big Bang Theory becomes increasingly preposterous with each new discovery and the failure to find any kind of limit to the universe. And why should there be a limit? How could it not go on forever and ever? You must be enamored with that last-gasp, Space-time stuff to believe there could be an end to the universe or that it was all wrapped

---

[14] http://go.glennborchardt.com/400billioninMW

[15] Conselice and others, 2016, The Evolution of Galaxy Number Density at z > 8; Smith, 2016, How many galaxies are in the universe?

[16] There is no agreement on the number of stars in the observable universe, although estimates tend to increase with improvements in instrumentation. Because the universe is infinite, conventional estimates are always too low and usually out of date. On 20160504, Wikipedia said that there are "approximately 170 billion (1.7×10^{11}) to 200 billion (2.0×10^{11}) galaxies" according to Gott and others (2005). As of 20161119, the new estimate was two trillion. Wikipedia says that the number of stars in the Milky Way "is estimated to be 200–400 billion" according to Frommert and Kronberg (2005). As mentioned above, more recent estimates use the 400 billion number. Assuming that the Milky Way is an average galaxy, that would mean that there are over $2 \times 10^{12} \times 4.0 \times 10^{11} = 8 \times 10^{23}$ stars in the observable universe—800 billion trillion.

up in a magical, 4-dimensional ball. It would make more sense if you assumed, instead, that the universe was 3-dimensional and not expanding. Then the astronomers' age estimate of 13.8 billion years for the most distant galaxies would yield an observed universe that was 27.6 billion light years across.

Now, get ready for an even more fantastic part. All that immensity, with over 400 billion trillion stars, is supposed to have exploded from a "singularity" smaller than the period at the end of this sentence. Once you assume something is expanding at a particular rate, you only have to project that rate backwards into the past until that something becomes nothing. Say you weighed 300 lbs and lost a pound a week. After 5.77 years, you would weigh nothing. That is how silly the Big Bang Theory is. It is a story of mathematicians gone wild.

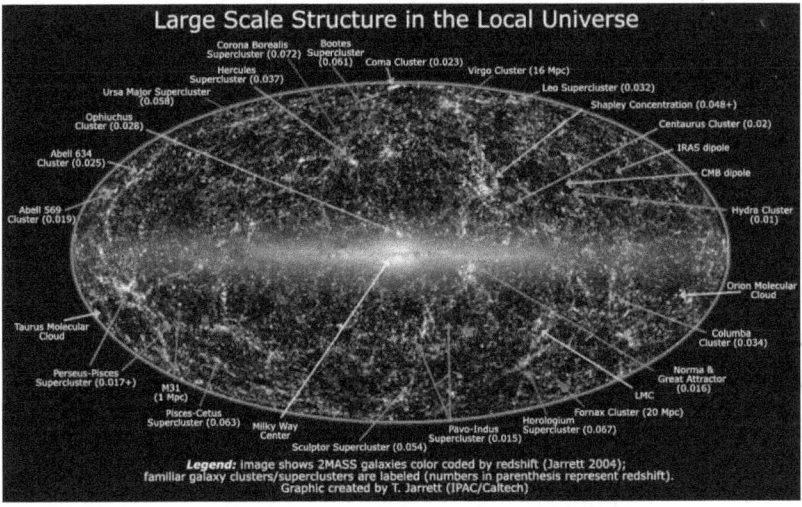

Figure 2 Image showing some of the 50,000 galaxies in the nearby universe detected by the Two Micron All Sky Survey (2MASS) in infrared light. Note that astronomers now estimate there are over two trillion galaxies, so this is less than 0.000003% of them. Credit: T.J. Jarrett (IPAC/Caltech).

Chapter 2

Galaxies in Collision

Figure 3 The Andromeda Galaxy. Credit: NASA.

For a finite universe, supposedly the product of an explosion, we are left with an interesting observation in M31, the Andromeda Galaxy, which is the closest spiral galaxy to our own Milky Way (Figure 3). It is so close you can even use binoculars to see it as a blur in the Andromeda constellation. That is only its center. With a telescope, its angular width on the sky appears to be at least 6 times the width of the Moon. The Andromeda Galaxy is special because its light is blueshifted, indicating it is coming toward us, unlike the redshifted light from most other galaxies. Galactic redshifts have many causes, but the most obvious for nearby objects involves the "Doppler Effect." On Earth, we observe the Doppler Effect whenever a train passes by as it whistles. When it is coming toward us, the pitch is high (short wavelengths); when it is going away from us, the pitch is low (long wavelengths). Light does the same.

Based on the amount of blueshift, astronomers figure Andromeda is coming toward us at over 100 km/s. This means it probably will join the Milky Way in about 4 billion years.[17] The question then arises: "If the universe is expanding and everything is coming apart, how come this huge object, a galaxy with a trillion stars, is coming toward us?" The conventional answer is that, in this special case, the force due to gravitational "attraction" is stronger than the force due to universal expansion. But collisions of galaxies appear to be relatively common (Figure 4). If the universe is expanding, it obviously is not expanding everywhere. In addition, it certainly is not as homogeneous as idealists believe it to have been right after the purported explosion. The heterogeneity we see around us is the result of a coming together, not a coming apart. The collisions of galaxies are just what would be expected in an Infinite Universe, while an expanding finite universe would have none (Figure 5).

Figure 4 Spiral galaxies NGC 2207 and IC 2163 in collision.

---

[17] Note that the cosmological ages I must use are from conventional sources, most of which were calculated by using the assumptions of the Big Bang Theory.

36 *Infinite Universe Theory*

Figure 5 An explosion has only divergence, while creation requires convergence. Credit: Vector Graphics.[18]

---

[18] https://go.glennborchardt.com/vectoreps

Chapter 3
Galaxy Clusters in Collision

At first, the fact galaxies sometimes collided despite the hypothesized universal expansion was of no particular concern to cosmogonists. These galaxies were, after all, parts of galaxy clusters, the next step up in the cosmological hierarchy. As such, they were supposed to be "gravitationally bound," which is what made them clusters in the first place. The galaxy clusters allegedly formed during the early universe. Then, as the expansion of the universe continued, the clusters supposedly separated from each other. Indeed, the distances between galaxy clusters are vast, with the nearest one, Virgo, being over 50 million light years from our own cluster, called the Local Group. Only one problem: many of them are not continuing to separate. Although the measurements are not very good, it seems our Local Group is approaching the Virgo cluster at something between 100 and 400 km/s. The discovery of the Bullet Cluster (Figure 6), as well as many other colliding galaxy clusters, confirms the "expanding universe" clearly is not expanding in all its parts. We are getting hints that at all scales, convergence is just as common as divergence in a universe that is only supposed to be diverging. The "force of expansion" must have been too weak to overcome the negligible "force of gravity" that supposedly brought these clusters together.

This tendency for galaxy clusters to merge appears to be common throughout the universe. Virgo has about 2000 galaxies, while the Local Group has about 50. Together, they form the Virgo supercluster, which in turn, is part of an even larger supercluster called Laniakea. Laniakea has at its center what is being called the "Great Attractor," which is tens of thousands more massive than the Milky Way. This is part of the hierarchical nature of the universe in which matter tends to form vortices centered on relatively large, dense nuclei.[19] In later chapters, we will discover why this phenomenon occurs and how gravitation and the formation of matter are linked.

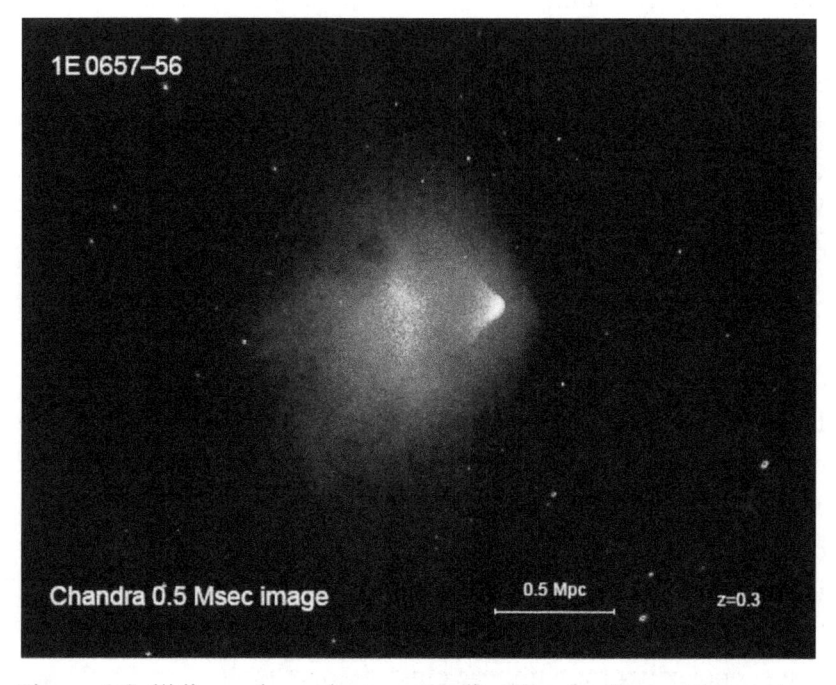

Figure 6 Colliding galaxy clusters. Credit: Chandra X-ray Observatory.

---

[19] Puetz and Borchardt, 2011, Universal Cycle Theory.

Chapter 4

Elderly Galaxies at the Edge of the Universe

Contradictions of the Big Bang Theory gather steam with each new discovery. According to the theory, we should be able to see what the early universe looked like by observing the most distant celestial objects. As we look further and further out into space, we should see younger and younger objects. For instance, we would not expect to see any galaxies with redshifts[20] correlative with the claimed 13.8-billion year age of the universe. If we saw whole galaxies at anywhere near that age, it would mean the Big Bang Theory is false. But lo and behold, in 2009, astronomers found just such an object near the outer reaches of the observed universe.[21] Its redshift was commensurate with an age of 12.8 billion years. Not only was it a full-fledged galaxy, but it also had a clearly identifiable "black hole" at its center—another indicator of great age. Of course, as one looks further and further into the Infinite Universe, cosmological objects take up smaller and smaller portions of the night sky. At the same time, they become dim and their redshifts increase. The redshift of that galaxy indicates the light traveled for 12.8 billion light years before reaching us. We are seeing the galaxy as it was 12.8 billion years ago, shortly after the universe supposedly exploded out of nothing 13.8 billion years ago. It would have been only a billion years old when that light left (13.8-12.8). Now, "black holes" form as the super-dense nuclei of galaxies *after* they begin to revolve— probably from glancing collisions with other galaxies. As you

---

[20] The redshift phenomenon will be discussed in numerous places scattered throughout the text. That is because, in addition to the Doppler Effect, it has many other causes and is critical to understanding both the Big Bang Theory and Infinite Universe Theory.
[21] Thompson, 2009, Most Distant Galaxy with Big Black Hole Discovered.

might expect, they take a long time to form. Our Milky Way Galaxy, considered 13.7 billion years old, contains a black hole. The solar system initially formed 4.5 billion years ago, but here is an entire galaxy supposedly formed in less than a billion years. In 2015, an even older galaxy was dated at 13.2 billion years.[22] This one supposedly formed in only 0.6 billion years. Unbelievable!

Normally, this discovery of an elderly galaxy at such a great distance would be enough to falsify the Big Bang Theory. There are numerous reasons the observation had little impact. For one thing, these kinds of "elderly" contradictions depended on improvements in instrumentation—long after the Big Bang Theory became an entrenched habit. For another, the Big Bang Theory currently is the only game in town, with the rules firmly established in modern physics—nearly everything in cosmology and physics is interpreted from that point of view. Startling observations contradicting such a major paradigm first need confirmation. Then the theory needs revamping to include the new observations under its umbrella. This takes time as well as an all-encompassing knowledge of the situation. Eventually, establishment theoreticians will find proper excuses such as "inflation" and "dark energy" that will keep the paradigm afloat. Practical astronomers who spend endless hours sitting on telescopes are unlikely to consider the ramifications of relativity and scientific philosophy. Theory is normally not their strong suit. Like engineers and earth scientists, they normally focus on their narrow specialty.

Nonetheless, contrary data keep piling up. The advent of the Hubble Space Telescope, launched into low Earth orbit (560 km) as part of the shuttle program, improved clarity and reduced light pollution such that extremely distant objects could be seen.

---

[22] Smith-Strickland, 2015, This is the Oldest Galaxy We've Found So Far.

Again, according to NASA and the cosmogonists, heavenly bodies should appear increasingly young as a function of distance (Figure 7). This is because the farthest bodies were supposedly formed shortly after the birth of the universe. Whether the universe is expanding or not, looking at distant objects is thus like looking back in time. By looking farther and farther back, one should see the young galaxies, then only stars, and then nothing. The first high-resolution Hubble photos showed some blurry objects at the maximum distance it could detect. These looked suspiciously like galaxies, but one could not be sure. Then, after solving some technical problems, detection improved and Hubble was able to obtain relatively clear photos of what is called the Hubble Ultra Deep Field (HUDF)—over 13 billion years into the past (according to the assumptions of the Big Bang Theory) (Figure 8).

And what a surprise! In whatever direction one looks, the HUDF photos show fully formed galaxies at distances only supposed to contain stars or, at most, juvenile galaxies (Figure 7). One photo shows over 10,000 galaxies in a tiny patch of sky (Figure 8). The galaxies in the HUDF look no different from those at shorter distances considered "normal" by NASA. Again, if the Big Bang Theory calculations were correct, some of the galaxies in the HUDF had to be only 400 million years old when the light was emitted. This cannot be. Galaxies cannot form in such a short time. After all, the fully formed galaxies they are seeing were over 12 billion years old when the light from them started its 13 billion-year journey toward the Hubble telescope. That means these objects must now be at least 25 billion years old. The upshot is that the unobserved portion in the outer reaches of the NASA diagram is pure conjecture. The youthful objects predicted by the Big Bang Theory will never be seen because, as plasma physicist Eric Lerner says "The Big Bang never happened."[23]

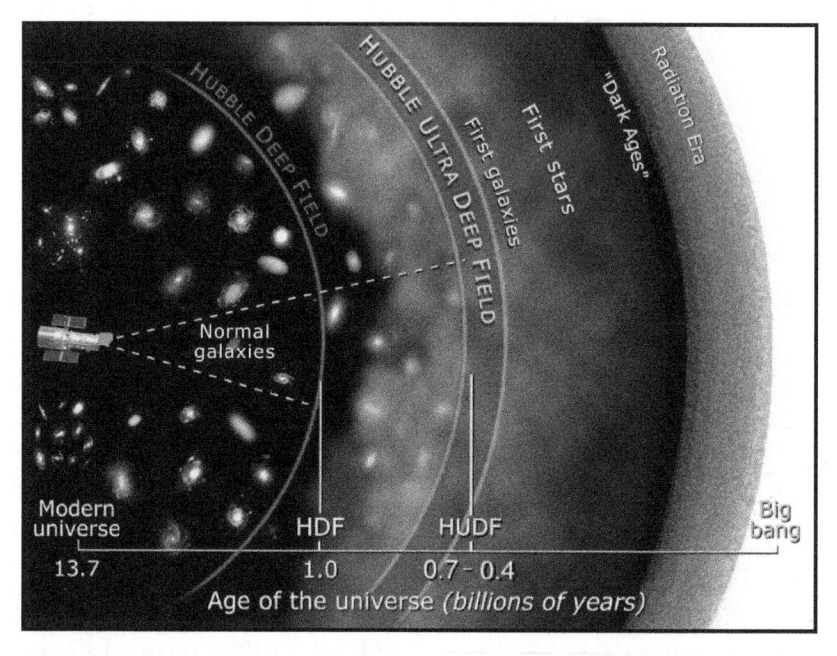

Figure 7 NASA's official view of what the Big Bang universe should look like (seriously).[24] Credit: NASA.

---

[23] Lerner, 1992, The Big Bang never happened.
[24] http://go.glennborchardt.com/NASABBT

Figure 8 There are over 10,000 galaxies seen in this photo of the Hubble Ultra Deep Field. It comprises a tiny patch on the sky 1/10 the size of the Moon. Credit: NASA, ESA, H. Teplitz and M. Rafelski (IPAC/Caltech), A. Koekemoer (STScI), R. Windhorst (Arizona State University), and Z. Levay (STScI).[25]

---

[25] http://go.glennborchardt.com/HUDFpic

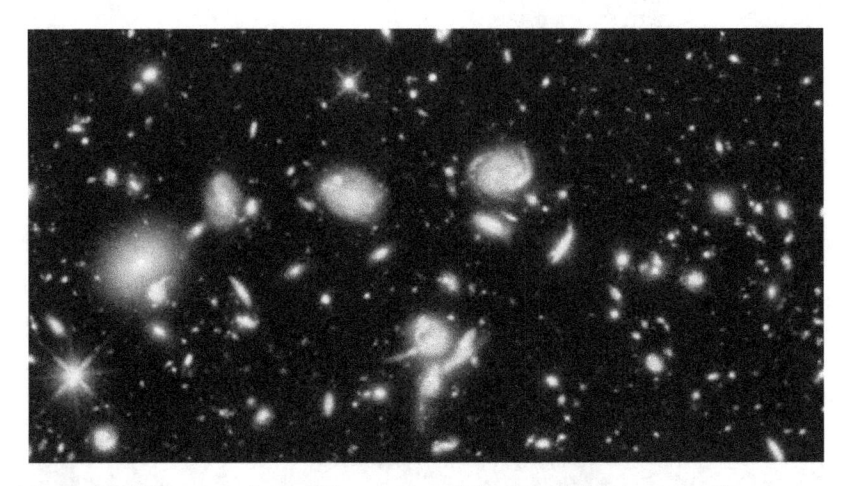

Figure 9 Close-up of a small portion of the HUDF. Note that these objects are various colors.[26] Most are not red as implied by the misnomer "cosmological redshift." <u>Color is determined by frequency, not wavelength</u>.

---

[26] http://go.glennborchardt.com/HUDFpic

Chapter 5

Elderly Galaxy Clusters at the Edge of the Universe

Individual galaxies form over vast regions of space. For instance, our Milky Way galaxy has a *diameter of about* 1.4×10^{18} *km*,[27] which is almost a million trillion miles (Figure 10). It is separated from the nearest galaxy, Andromeda, by about 2.5 million light years—about 17 galactic diameters away from us. The main point: Galaxies apparently require such huge portions of the universe for them to form independently from other portions. The Milky Way had to be gravitationally isolated from other galaxies for at least 13.7 billion years. It will take at least another 4 billion years for both galaxies to "collide," forming a relatively young aggregation containing 17.7-billion year old stars.

Obviously, galaxies must form first before they can join with other galaxies to form a cluster. It is not clear how long this initial formation and subsequent agglomerating process might take. The Milky Way and Andromeda already are part of the cluster we call the Local Group. We do know that, at an estimated approach velocity of 100-400 km/s it would take 20 to 80 billion years for our 50-galaxy Local Group to join the 2000-galaxy Virgo cluster. As generally happens when things come together, clusters develop emergent characteristics. Well-developed galaxy clusters contain a massive nucleus filled with an *intracluster medium* consisting of plasma of hydrogen and helium nuclei that is extremely hot—at least 10 million degrees Kelvin. Presumably, this is where whole galaxies go to die as they gravitate toward that medium. Unfortunately, until a timeline is developed on the evolution of the medium, we must

---

[27] http://go.glennborchardt.com/MWrippling

accept redshift ages of the remaining galaxies as representative of the age of the clusters.

Nonetheless, constituent ages are instructive and seem to be getting exponentially older by the second. In 2005 the farthest cluster observed was 9 billion light years from us.[28] In 2010 the farthest cluster observed was 9.6 billion light years from us.[29] In 2011 the farthest cluster observed was 10.7 billion light years from us.[30] In 2014 the farthest cluster observed was 13.3 billion light years from us.[31] Remember the oldest independent galaxy was 13.2 billion years old, so none of this accounts for cluster formation time, which supposedly occurs because of gravitational attraction. You can see where this is going. Soon, the race to find the oldest galaxy will find one older than the universe itself! Then, like Lucy, the cosmogonists will have some more 'splaining to do.

Here are some details on the 9.6 billion-year old galaxy cluster, CLG J02182-05102, which is dominated by old, red and massive galaxies typical of present-day, nearby clusters. For example, it is similar to a young version of the Coma Cluster we observe today, which supposedly had billions of more years to develop.

If the Big Bang Theory was true, a galaxy cluster at a distance of 9.6 billion light years would be only 4.2 billion years old when it emitted the observed light (13.8-9.6). After that little colliding galaxy cluster incident, some cosmogonists changed their minds about clusters having formed during the early universe. Galaxy clusters apparently have many different ages. In fact, one of them is surprisingly young. It supposedly formed in

---

[28] Wanjek and Steigerwald, 2005, Most Distant Galaxy Cluster Shows Universe 'Grew Up Fast'.
[29] Hadhazy, 2010, Ancient City of Galaxies Looks Surprisingly Modern.
[30] Gary, 2011, Astronomers find old heads in a young crowd.
[31] Vergano, 2014, Hubble Reveals Universe's Oldest Galaxies.

only a billion years, since it is 100 times less massive and produces new stars at 100 times that of other nearby galaxy clusters.[32] Our own galaxy cluster, called the Local Group, is at least 13.7 billion years old, judging by the age of one of its constituents, the Milky Way. Unfortunately, for the Big Bang Theory, many of the galaxies in the supposedly 9.6 billion-year old cluster look a lot older than that. Again, for the Big Bang Theory to be correct, galaxies and their clusters should look younger and younger as we look farther and farther out. That is why it was such a shock when astronomers found these elderly galaxy clusters containing elderly galaxies. It would be like finding a teenager in your bassinet. Again, if the universe did explode out of nothing, the galaxies at great distances would be babies as shown in NASA's illustration (Figure 7), but they are not.[33] Philosophers of science claim such disproofs or "falsifications" are supposed to lead to theory rejection.

All this is evidence that galaxies, and now galaxy clusters, in the supposed distant "early universe" of the Big Bang Theory are similar to the galaxies and galaxy clusters close to us. This is what we would expect if there were no Big Bang. The above ages proffered by the cosmogonists cannot possibly be correct. Many of those galaxies are just like our own 13.7 billion-year old Milky Way galaxy. If they are at a redshift distance commensurate with an age of 13.2 billion years, they would have to be over 26.9 billion years old now. The cosmological redshift may be an adequate measure of cosmic distance, but it certainly is not a measure of cosmic time, as I will explain in more detail later.

---

[32] Anon, 2005, Galaxy Clusters Formed Early.
[33] Hadhazy, ibid.

48   *Infinite Universe Theory*

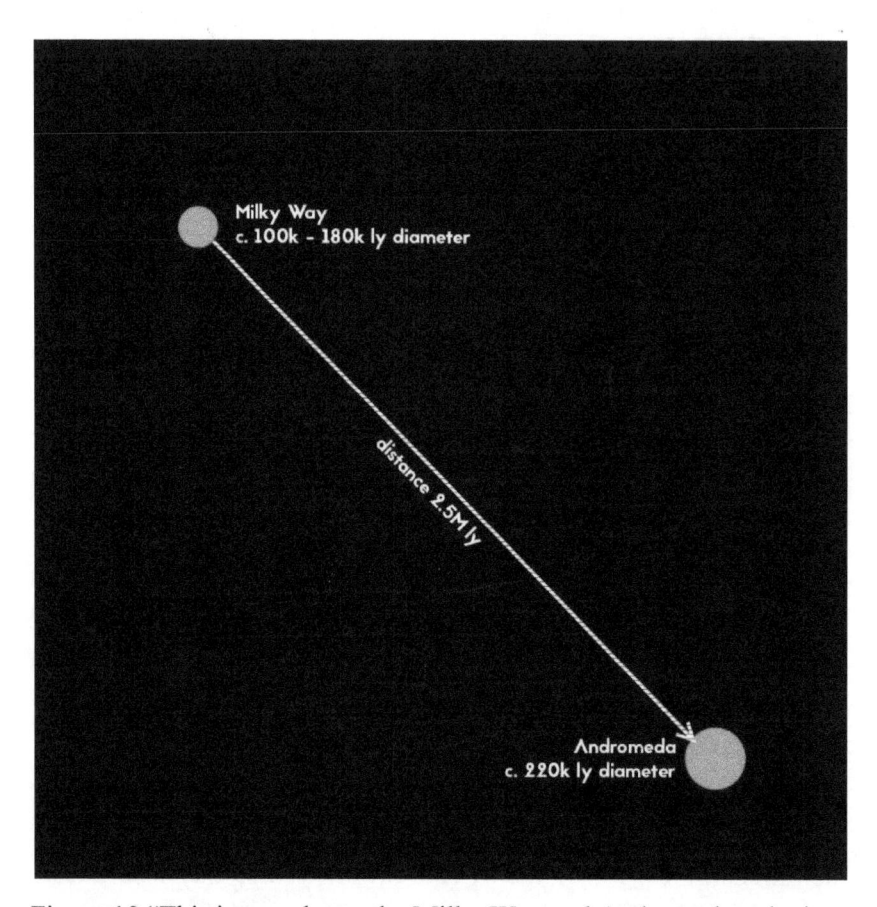

[Figure: Diagram showing Milky Way (c. 100k - 180k ly diameter) and Andromeda (c. 220k ly diameter) galaxies with distance 2.5M ly between them, to scale.]

Figure 10 "This image shows the Milky Way and Andromeda galaxies in space to scale. It illustrates both the size of each galaxy and the distance between the two galaxies…" ly = light year = 9 trillion km = 5.58 trillion miles.[34] Credit: Jan van der Crabben.[35]

---

[34] http://go.glennborchardt.com/MilkyWayandAndromeda
[35] (Own work) [CC BY-SA 4.0 (https://creativecommons.org/licenses/by-sa/4.0)], via Wikimedia Commons.

Chapter 6
Solipsism and Perception

We are born at the center of the universe. Cry and the universe feeds us. Make a mess, and the universe cleans it up. By the time we are a year old, we can put a blanket over our heads and make the universe go away. The self-absorption we are born with is a philosophical tendency called "solipsism," the belief the existence of the universe depends entirely on us. People naturally follow the "Principal of Least Effort," with limited powers of observation allowing us to focus only on the things and activities nearby.

And so it was that, during the 20th century, the entire human race was taking its last few steps out the incubator, shedding the remains of what is surely a natural, but misbegotten philosophy. This was to be expected. It was expected in the form of "reverse recapitulation, in which the phylogeny of the group retraces the ontogeny of the individual." What that means in plain language is that, like all microcosms, many aspects of the evolution of the group are similar to the evolution of the individual.

During the first part of its evolution, humanity studied its immediate surroundings in the same way the infant first studies the nearest objects in its universe. Those who looked the farthest, the cosmologists, nevertheless were born with the same solipsistic tendencies. As each instrument allowed us to see further, we expanded our interpretation of just what we were seeing. Early on, we thought the universe consisted of two spheres: the stationary Earth surrounded by a transparent rotating sphere with attached stars—a special place created just for us. The Copernican Revolution said no, it was not Earth that was stationary, but the Sun, which was at the center of the universe. Still later, the revelation that the Sun was just an ordinary star among the 400 billion within the Milky Way galaxy must have been yet another shock to solipsism.

The Milky Way contained all the celestial objects we could see. Or did it? Some distant fuzzy objects were recognized as early as the 10th century. In the 18th century, Kant speculated that these were independent star systems, using the term "world systems" like the Milky Way to describe them.[36] Confirmation of that step out of the solipsistic box did not occur until the 20th century. In 1912, Vesto Slipher was innocently looking at spectra from spiral galaxies to see if they contained chemical elements similar to our planetary system. Instead, he found that light from the Andromeda "nebula" was blueshifted.[37] A few years later, he showed light spectra from 12 out of 15 galaxies were redshifted, concluding those objects were moving away from us due to the Doppler Effect. In 1922, Ernst Opik made distance measurements proving Andromeda was outside the Milky Way. By 1929, Edwin Hubble discovered enough galaxies to establish that the degree of redshift also was a rough inverse function of their luminosity, which is a measure of their distance (Figure 11). His greatest mistake was to promote Slipher's speculation that this cosmological redshift always was due to the Doppler Effect. The title of his famous 1929 introductory paper says it all: "A relation between distance and radial velocity among extra-galactic nebulae."[38] Despite that grand entrance, he ended the paper with this widely ignored caveat showing that, unlike Slipher, he was seeking a more rational explanation for the redshift-distance correlation:

> *The outstanding feature, however, is the possibility that the velocity-distance relation may represent the de Sitter effect... In the de Sitter cosmology, displacements of the spectra arise from two sources, an apparent slowing down*

---

[36] Kant, 1755, Universal Natural History and Theory of the Heavens, p. 43. [Note: I could not find the oxymoronic term "island universe" in his work, although it is often attributed to him.]
[37] Slipher, 1913, The Radial Velocity of the Andromeda Nebula.
[38] Hubble, 1929.

of atomic vibrations and a general tendency of material particles to scatter.[39]

Although the de Sitter effect was never accepted as an alternative, this quote shows that Hubble also was entertaining the "tired light effect" to avoid the solipsistic absurdity that each of the galaxies were rushing away from us at once.

By 1947, Hubble said:

If they are valid, it seems likely that red-shifts may not be due to an expanding universe, and much of the current speculation on the structure of the universe may require re-examination.[40]

This was nearly 20 years after his discovery precipitated the universal expansion craze and right before the Big Bang Theory became popular. Despite the talk about the Hubble "expansion," he never fully endorsed that explanation.[41] In his last lecture, Hubble said that without certain fudge factors that assume recession, "the law will represent approximately a linear relation between red-shifts and distance."[42] Unfortunately, like Eisenhower's caveat about the dangers of the military industrial complex and Einstein's recantation, Hubble's belated warning went unnoticed by the establishment. There was a big reason for that having to do with the philosophical backlash physics was undergoing at the time. This was so pronounced that I have termed the result regressive physics.

The backlash was in keeping with our traditional solipsistic and myopic tendency to keep ourselves at the center of everything. The fact that the observed universe is spherical,

---

[39] Ibid, p. 173.
[40] Hubble, 1947, The 200-inch telescope and some problems it may solve, p. 165.
[41] Sauvé, 2016, Edwin Hubble... and the myth that he discovered an expanding universe.
[42] Hubble, 1953, The Law of Red-Shifts, p. 666.

52   *Infinite Universe Theory*

being 13.8 billion light years in every direction, only means that what we are seeing is simply a matter of perception within the Infinite Universe. Improvements in instruments will widen the limit of perception, but the Big Bang creationists will persist in making up ad hoc stories to mollify their religious patrons.

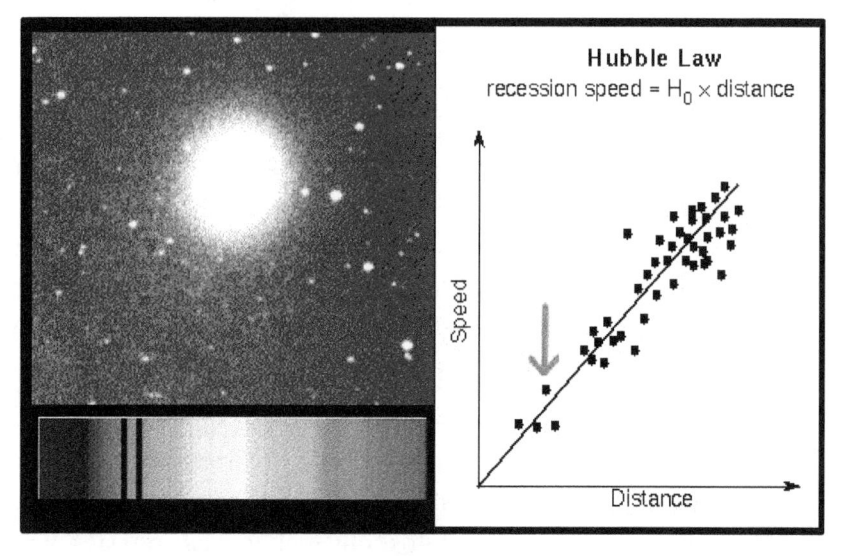

Figure 11 Typical redshift vs. distance plots calculated as erroneously assumed recessional velocities. This is part of an animation prepared by the Institute for Astrophysics and Space Science, Western Kentucky University.[43]

---

[43] http://go.glennborchardt.com/Hubble-Law

Chapter 7
Einstein's "Untired Light Theory"

As we saw, Hubble was not immune to the philosophical surge toward indeterminism that became fashionable during his heyday. Hedging his bets in 1936, he wrote:

> The interpretation is consistent with the data whether the nebulae are pictured as scattered through an old-fashioned, infinite universe or whether a homogeneous model obeying relativistic laws of gravitation is adopted...[44]

He continued in this ambivalent vein to suggest the alternative to galactic recession might be due to "a hitherto unrecognized principle of nature."[45]

As always, agnosticism prevents clear thinking. This one left an opening for cosmogonists who were then able to use his data in support of their misinterpretation that the universe was expanding. That is not unusual—it is typical of the determinism-indeterminism philosophical struggle. Life's struggles require weapons for both defense and offense. Publications and ideas are like swords that can be used in philosophical battles. That is why religious books are big sellers among believers and why atheistic books are big sellers among nonbelievers. It is why Darwin's "Origin" became an instant best seller. Capitalists no doubt used it in their skirmishes with the remnants of feudalism that hindered their progress. Hubble's 1929 paper came along as the philosophical scrubbing of physics was well underway—materialism and communism was to be conquered at last. Despite philosophical errors and dubious math, relativity has been "proven right" on most every occasion. Here was just another circumstance for an indeterministic interpretation to rise to the fore.

---

[44] Hubble, 1936, Effects of Red Shifts on the Distribution of Nebulae, p. 542.
[45] Ibid, p. 553.

Normally, it would be strange for scientists to choose the Doppler Effect as the mechanism for the redshift phenomenon. The Doppler Effect describes wave motion in a medium. It could not possibly occur in Einstein's assumed perfectly empty space. For instance, the example I gave for the Doppler Effect for sound would not work on the Moon. This is because the Moon has no atmosphere—the medium for sound waves is absent. However, since the young Einstein's dismissal of aether, there was not supposed to be a medium in outer space. How could there still be a Doppler Effect? Also, according to Einstein, light was a massless particle that could travel through perfectly empty space without losing energy. These paradoxes became prominent when Einstein introduced his eight ad hocs in support of his corpuscular theory of light. I will discuss these in detail in Chapter 15.1.

Despite his doubts, Hubble apparently adopted the aether denial necessary for that view of light. That is probably why he was a bit naïve in saying that the redshift might be due to "a hitherto unrecognized principle of nature." At that time, scientists already knew what the principle was: The Second Law of Thermodynamics. That law essentially says that perpetual motion is impossible—there has never been an exception. That is why the patent office rejects such claims outright. In media, such as water, waves get longer and longer as they travel great distances, necessarily losing some of their motion in the process (Figure 12).

Figure 12 Ripple wavelength increases with distance. Credit: Exo.net.

The upshot is that no thing or motion in the universe can go from point A to point B without something happening to it, thereby decreasing some aspect of its motion. The opposite point of view, since used to support the universal expansion hypothesis, is what I call the "Untired Light Theory." It is especially ironic that any attempt to get a patent based on this flagrant violation of the Second Law of Thermodynamics would be rejected forthright by Einstein himself when he worked at the Swiss patent office. In a relatively homogeneous medium, any increase in wavelength, as in the cosmological redshift, would mean that some energy has been lost to the macrocosm. Each wave involves a convergence and a divergence that produces the next wave. The next wave is similar to the last one, but it is never identical. What does change is the slight decrease in the ability of a wave to produce the next wave. Eventually, waves spread out from the source, being reproduced in a form not quite as true as the last (Figure 12). This is analogous to the old "telephone chain" and to the reproduction of the cells in your body. Eventually, you will not look the same anymore. This is the reason immortality is physically impossible. Think of the cosmological redshift in the same way.

The "Untired Light Theory" is completely idealistic, unproven in practice, and even unpatentable. It is obvious that there *are* energy losses. That is exactly what the redshifts show: long waves have less energy than short waves. There still would be Doppler effects especially noticeable at short distances (such as Andromeda), but the "tired light effect" would dominate when the distance is especially great. Light, of course, clearly has wave properties such as frequency, refraction, diffraction, etc. Astonishingly, physicists handled the contradiction by regarding the photon as both a particle and a wave at the same time: a wave packet. This is our first inkling that Einstein and his Special Relativity Theory are inextricably tied to the Big Bang Theory.

Redshifts produced due to tired light would be a function of distance, just as Hubble showed with his measurements on distant galaxies. However, in adopting Einstein's corpuscular theory of light, idealists insisted that the hypothesized photons traveling through perfectly empty space had no reason to lose energy—despite the Second Law of Thermodynamics. Even this does not make sense, for space still could not be perfectly empty if photons were traveling through it. These corpuscles traveling from light sources at various angles would inevitably crash together, losing motion per Newton's Second and Third Laws of Motion. None of that happens because photons are purely imaginary. That is why they have no rest mass, are central to the "Untired Light Theory," and do not obey the Second Law of Thermodynamics. They do not and cannot exist.

Because redshifts are relatively easy to measure, data accumulated rapidly, showing conclusively that redshifts and brightness of galaxies were inversely correlated. That is, the higher the redshift, the dimmer the galaxy, and the further away it was from us as we saw in Figure 11. It then was easy for mathematicians like Hawking to back-calculate to the time when the expansion must have started from a single point. This "age of the universe" necessarily has changed from time to time as instruments, data accuracy, and knowledge about the distance relationship improved. For instance, in 1949, George Gamow, one of the instigators of the Big Bang Theory, informally estimated the universe to be only *3 billion years old*.[46] That was a gross under-estimation, with the calculated age creeping upwards ever since. By 1970, the solar system alone was confirmed to be over 4.5 billion years old.[47] Now there seems to be a temporary consensus among cosmogonists that the universe is 13.7 billion years old—no, wait, they just changed it to *13.8 billion*, claiming that "In far less time than it takes to blink an eye, the universe blew up by 100 trillion trillion times in size."[48]

That may involve some elegant mathematics, but the resulting delusion is built upon indeterministic assumptions. You can dismiss that by holding fast to one deterministic assumption in particular: ***infinity***. Of course, the physicists and cosmogonists responsible for this most recent embarrassment hold just as fast to the opposite assumption: *finity*. Changing their minds about the nature of the universe will involve a long, hard struggle. Stay tuned for my prediction about when that will occur and what it will take to accomplish it. In the meantime, be sure to keep a copy of the latest *Big Bang propaganda*[49] as a collector's item.

---

[46] http://go.glennborchardt.com/Gamov3Ga
[47] Tatsumoto and Rosholt, 1970, Age of the Moon.
[48] Harrington and Clavin, 2013, Planck Mission Brings Universe into Sharp Focus.
[49] http://go.glennborchardt.com/BBTad

handwritten: 2,000,000,000,000 ↑ 1000 Galaxies for each 8 people on our world!

Chapter 8

Space-time Salvation

Initially, there was another particularly galling problem with the expanding universe hypothesis required for the Big Bang Theory. Wherever one looked, north, south, west, east, up, or down, the maximum redshift was correlative with what is now considered 13.8 billion years. How could Earth be at the point where the Big Bang began? No doubt, a few old-fashioned religiously inclined solipsists thought it was humanity's pre-Copernican special privilege to be at the center of all things. Slightly cooler heads prevailed. There were just too many galaxies for that notion, even then. As mentioned, current estimates as to the number of observable galaxies are about two trillion and counting. The upshot is that the chance of our galaxy being at or near the origin of the Big Bang in a 3-D universe also is about one in two trillion. In addition, we are on the margins of our own galaxy. Our Sun is nothing special among the 400 billion stars in the Milky Way.

Faced with such astronomical odds, cosmogonists pulled out their last remaining ace: General Relativity Theory. In 1916, Einstein treated time mathematically as an extra dimension.[50] Now, in math one can treat any particular factor as a dimension—as long as you do not know what a dimension is, and in this case, if you do not know what time is. In Special Relativity Theory, this ignorance allowed Einstein to assume incorrectly that length and time were equivalent.[51] That would be like assuming frogs and jumping were equivalent. Nonetheless, in 1919, Eddington's eclipse observations were widely promoted as confirmation of Einstein's prediction that gravitation was due

---

[50] Einstein, 1916, The foundation of the general theory of relativity.
[51] Bryant and Borchardt, 2011, Failure of the relativistic hypercone derivation.

to space-time curvature.[52] In 1922, Alexander Friedmann of the Soviet Union provided the mathematical details necessary for a 4-dimensional universe.[53] Of course, the interpretation that most galaxies were receding from us fit right in.

The fantastic conclusion that we live in a 4-dimensional universe still creates skepticism among those with a bit of common sense. All real things we know have three dimensions, xyz. Nonetheless, cosmogonists hold fast to their belief in the existence of space-time. Without space-time, the Big Bang Theory would be dead. Because space-time is critical to the Big Bang Theory, I will present a detailed analysis after you read the chapter on neomechanics.

In the rest of the book, I will attempt to show you that Infinite Universe Theory is not that difficult to understand. Sure, there is some math and simple physics involved, but it is not much and I think you will find those changes beneficial, resulting in a theory without the contradictions invariably found in the Big Bang Theory (Table 4).

---

[52] Dyson and others, 1920, A Determination of the Deflection of Light by the Sun's Gravitational Field.
[53] Friedman, 1922, Über die Krümmung des Raumes.

Table 4 Falsifications, contradictions, and paradoxes disproving the Big Bang Theory.

The Big Bang Theory predicts that we should observe only young cosmological objects at great distances. Instead, we see elderly galaxies and galaxy clusters at the limit of observation.
Cosmological objects often collide. In actual explosions, objects are scattered in all directions and do not collide.
The opinion that the universe is expanding is dependent on the "Untired Light Theory," which assumes that light can travel great distances without losing energy. Nothing we know of can travel from one place to another without losing energy.
The explosion of the universe out of nothing is a contradiction of the First Law of Thermodynamics, otherwise known as the conservation of energy.
The Doppler Effect, considered responsible for the Cosmological redshift and the interpretation that most galaxies are receding from us, only occurs in a medium. Einstein's corpuscular theory of light, denies the necessary presence of a medium.
Einstein's **objectification** of time is invalid. Time is not an object; time is motion. The space-time concept, as used in General Relativity Theory and Big Bang Theory, assumes time to be a dimension, which it is not. The universe is 3-dimensional, just like everything we observe. "Time dilation" and other Einsteinian fantasies are products of aether denial.
The Big Bang Theory is based on the assumption of *finity*. The most plausible assumption is *infinity*. There are over two trillion galaxies in the observable universe with no end in site. An Infinite Universe cannot expand, because it is already full.
The existence of the universe implies nonexistence (empty space) is impossible. There is no definitive evidence for perfectly empty space. *Infinity* implies the existence of aether.

//
PART II: INFINITE UNIVERSE THEORY

Infinite Universe Theory is the ultimate theory of the cosmos. The reason is simple. There is nowhere else to go. How could there be an end to the universe? Alternatives such as the steady state[54] and multiverse[55] theories are mere reformist stopgaps founded on the same erroneous assumptions that brought us the Big Bang Theory in the first place. Proponents of a financially successful paradigm generally do not agree that there could be anything "erroneous" about their founding assumptions. In fact, most would not admit that their work is based on any particular assumptions at all. Nonetheless, all theories are founded on fundamental assumptions.

The shift from one theory to another always requires a shift in assumptions. A revolution occurs when we adopt opposing assumptions. Thus, when Copernicus assumed that the Sun, rather than the Earth was the center of the solar system, he produced a revolution in cosmology. When Wegner assumed that continents were in motion instead of being fixed, he produced a revolution in geology. The shift from the Big Bang Theory to the Infinite Universe Theory requires a similar, but more comprehensive shift in assumptions. So far, you have read a few complaints about the Big Bang Theory, but I have not yet proposed a viable alternative. If you are skeptical about this, I do not blame you. Abandoning a decrepit ship is unwise unless a better one is available. If you are practical and logical, you will find Infinite Universe Theory to be much more satisfying than the current Sci-fi fantasy. Unfortunately, the shift from old to new is rather difficult for most folks. It requires a basic understanding of the logical foundations of both the old and the

---

[54] Bondi and Gold, 1948, The steady-state theory of the expanding universe; Hoyle, 1948, A new model for the expanding universe.
[55] Hawking and Mlodinow, 2012, The grand design; Kashlinsky and others, 2010, A New Measurement of the Bulk Flow of X-Ray Luminous Clusters of Galaxies.

new. None of this involves much math, but it does require one to be receptive to new ideas.

What makes this especially tough is the fact that the struggle between the Infinite Universe Theory and Big Bang Theory is part of the philosophical struggle between determinism and indeterminism. Throughout history, indeterminism has underpinned traditional beliefs that promise heavenly salvation from the tribulations found on Earth. Scientific advances were accepted, but only reluctantly after they proved their economic and political worth. Infinite Universe Theory has been so long in coming because any version of cosmogony has been seen as preferable to it. We are stuck with a universe exploding out of nothing because it is currently the most expedient compromise between science and religion. Most folks demur from second guessing physicists and cosmogonists otherwise advertised as the smartest folks around.

You live at a unique juncture in history. It is a short interval within which we can reject many of the falsehoods that surround us by rejecting a single assumption: *finity*. Someday, we will consider the Infinite Universe Theory as matter of fact. About the Big Bang Theory, we will ask: "What were they thinking?" This book will help you understand how and why the Big Bang Theory and its fantasies got as far as they did. Why was the explosion of everything out of nothing, the expanding 4-dimensional universe, complete with wormholes, and dilating time taken seriously? The story unfolds in the first half of the 20th century when physics was in crisis and the whole world was at war. Physics could have taken one of two paths: deterministic or indeterministic. For sociological reasons to be explained later, the indeterministic path was taken. Physics has been the worse for it ever since.

This regression in physics led by Einstein must be reversed. Physics must be revamped, not simply by changing basic equations, but by ridding ourselves of their indeterministic interpretations. I do this by laying the assumptive foundation and by proposing a new way of looking at mechanics, which I call "neomechanics." Even if you never learned a stitch of physics, I think that your study of neomechanics will be worth your time. Without it, I do not think you would be able to understand Infinite Universe Theory. I will try to make neomechanics as clear as possible, but first let us summarize the assumptions upon which neomechanics and Infinite Universe Theory must be built.

Chapter 9
The Ten Assumptions of Science

Although you might read some of the billions of pages defending the Big Bang Theory, you probably will not come across an argument like this. That is because a paradigm shift of such historic importance requires a radical change in assumptions. Such an about-face will not be easy, because most of us grew up with assumptions that support the Big Bang Theory. To accept the opposing assumptions I am about to present, you once again must have an open mind. Otherwise, you will remain stuck with four dimensions and a universe exploding out of nothing long after the younger generation has discarded those moribund ideas.

Progress sometimes means changing our minds. However, mental pathways, once established, remain so as a matter of habit. The more we use them, the more ingrained they become. As Silicon Valley knows, thinking "outside the box" is for the young, not the elderly. Change is especially difficult when we must admit that we were wrong. As scientists, our occupation is dependent on our being right. Not too many employers will hire us to be wrong. After decades of promotion by prestigious scientists and gullible media, any disavowal of cosmogony must lead to across-the-board embarrassment. As implied above, this short summary of what I call "The Ten Assumptions of Science" is necessary for understanding Infinite Universe Theory. You may find it ironic that, because the universe is infinite, we can never provide a final, complete proof that any of these assumptions are true. Nevertheless, we can still accept them as true without some kind of magical final proof. Like axioms in math, these are the beginning points for the logic to follow. In reading some of these, you may get the idea that they are self-evident, sort of like Kant's "a priori," which he mistakenly thought to be indisputable and independent of experience with the outside world.

The reason is that, what may seem self-evident to you may not seem self-evident to others. That is because each person's criteria for "self-evidency" are dependent on the chain of causal events they experienced. If that were not the case, if each of us had the same experience, there would be no philosophical disagreements and no need for this chapter. Everyone would be born with the memo that the universe is infinite.

I first enunciated the "ten assumptions of science" in the 1984 review manuscript,[56] published the complete version as a short book in 2004,[57] and used them as the foundation for "The Scientific Worldview" in 2007,[58] and "Universal Cycle Theory" with Stephen Puetz in 2011.[59] The summary below does not do them justice. If this is new to you, I recommend reading the detailed arguments in "The Ten Assumptions of Science" or chapter 3 in "The Scientific Worldview." If you already did that, this should be a good review. The assumptions are more or less in order of importance. First, we assume that matter exists. Second, we assume that all matter is in motion. Third, we make numerous assumptions about what we, as observers, can learn about those two fundamental phenomena.

Halton Arp, one of the most famous opponents of the Big Bang Theory, rebuffed my entreaties for him to read "The Ten Assumptions of Science."[60] In this, Halton supposedly was following in the footsteps of Newton, who is quoted repeatedly as having written: "Hypotheses non fingo" ("I feign no hypotheses"). The complete quote is instructive:

---

[56] Borchardt, 1984, The Scientific Worldview.
[57] Borchardt, 2004, The Ten Assumptions of Science.
[58] Borchardt, 2007, The Scientific Worldview.
[59] Puetz and Borchardt, 2011, Universal Cycle Theory.
[60] An email exchange shortly after Borchardt (2004) was published.

> *I have not as yet been able to discover the reason for these properties of gravity from phenomena, and I do not feign hypotheses. For whatever is not deduced from the phenomena must be called a hypothesis; and hypotheses, whether metaphysical or physical, or based on occult qualities, or mechanical, have no place in experimental philosophy. In this philosophy, particular propositions are inferred from the phenomena, and afterwards rendered general by induction.*[61]

This is representative of the empirical point of view, which, as Arp revealed, is still prevalent today, at least with regard to fundamental assumptions. Empiricists usually claim to be purely objective and, like Arp, they usually are unaware of the fundamental assumptions necessary for their investigations. Newton's outburst was a reflection of his frustration at being unable to discover the physical cause of gravitation. Hypothesis testing, of course, is a major part of science. We seldom do experiments without having some inkling of how they might turn out. We gather such data to support or falsify (disprove) a supposition, assumption, or hypothesis. Hardly anyone wants to repeat experiments in support of hypotheses that already have been proven or disproven. Journal editors, in particular, would not be impressed. Newton's standpoint is particularly ironic in that his work was mostly hypothetical rather than experimental. He actually feigned plenty of hypotheses. In fact, he later came up with one for the physical cause of gravity,[62] which was similar to the one we discovered independently and proposed in "Universal Cycle Theory"[63] and our paper on Neomechanical Gravitation Theory.[64]

---

[61] Newton, 1726, Philosophiae Naturalis Principia Mathematica, p. 943.
[62] Newton, 1718, Opticks, p. 325
[63] Puetz and Borchardt, 2011, Universal cycle theory.
[64] Borchardt and Puetz, 2012, Neomechanical gravitation theory.

The truth is that pure empiricism is of little value. One needs some reason (i.e., "hypothesis") to go to the trouble of gathering data. Measuring every pebble and sidewalk crack in the world would be a waste of time. At a minimum, we need preconceived notions or subconscious presuppositions to begin the work of science. None of these presuppositions just pop into our heads from nowhere. Again, every one of them is the result of our previous experiences whether we realize it or not. That is why we often can predict the beliefs of people simply by knowing something about their backgrounds: where they grew up, who their parents were, where they went to school, who their associates were, which groups they belong to, etc.

I did not realize how critical this was until I read the noted idealist R.G. Collingwood's "Essay on Metaphysics."[65] In the old days, we used to roam what they called "stacks" in libraries. I passed up the book on three occasions before opening it. Before that, I thought of "metaphysics" as mystical hocus-pocus of no use in science. Collingwood taught that which "goes beyond physics" could be either nonphysical or physical. The nonphysical stuff indeed was hocus-pocus, but I quickly grasped onto the idea that what "goes beyond physics" is simply more physics. No wonder I got into ***infinity***.

While there are causes for phenomena, there are causes for hypotheses too. The same folks who are enamored of the concept of "free will" generally are unaware of the causal chain that produces their belief. The causal chain stems, of course, from our traditions. If you want to understand why people believe as they do, you must study their past. Your religious beliefs, for instance, stem from where you happen to be born.

---

[65] Collingwood, 1940, An essay on metaphysics.

I was in the U.C. Berkeley library to discover why it was that exceptionally smart people, especially those at the best public university in the world, thought there were more than three dimensions and that the universe exploded out of nothing. Collingwood repeatedly emphasized that all thinking begins with presuppositions, subconscious assumptions that we do not realize we have. According to Collingwood, metaphysics was the discovery of what these presuppositions were. Once you bring a presupposition into the light of day through the spoken or written word, it becomes an assumption. From thenceforth you can study, discuss, and test it just like any other object. Of course, we use presuppositions or assumptions every day. We could not take another step without supposing that the ground beneath our feet would support us; we could not drive down a two-lane road without assuming that oncoming vehicles will stay in their lane. Those who claim that the practice of assuming is wrong are wrong.

But how was I to find out which assumptions were so fundamental that they could be used to discover why I thought differently from the recognized experts? Collingwood had a plan for that.

The first part of the plan was that, to be fundamental, an assumption must have an opposite, a contradictory claim that was false if the first was true, and *vice versa*. As an example, one might assume that the universe is infinite or finite. They are mutually exclusive, despite the illogical special pleading of those who cannot make up their minds. The universe cannot be 95% infinite and 5% finite. It either is or is not. Of course, when you finally decide which of the two fundamental assumptions to use, you will be taking sides in the philosophical struggle. You will be leaving the crowded ranks of the agnostics who are confused, have not studied the issue, or are more interested in popularity than logic. The fact that there are agnostics and folks opposed to a particular proposition is a clue to its status as a fundamental assumption. If everyone agreed with a proposition, we probably would not consider it an assumption.

The second part of Collingwood's plan was that, to be fundamental, neither of the opposed assumptions could be proven completely true without a doubt. This is what Karl Popper, the famous philosopher of science, was alluding to when he said that scientific theories could not be proven, although they could be either supported or falsified.[66] Again, as an example, no one is going to be able to go to the edge of the universe to find out whether it is infinite or finite. Of course, as you will see later, both Collingwood and Popper made these claims based on an assumption they did not acknowledge: infinity.

The third part of the plan was that, if you wished to consider more than one fundamental assumption as true, then both must be consupponible. This is an excellent, if unpopular word—I never saw it in a dictionary. One might use other words, such as "non-contradictory" or "coherent," but this means "with supposition," which is exactly what Collingwood was trying to get across. Not having contradictions and having coherence is nice, but those words do not necessarily imply that suppositions are involved.

Thus, armed with Collingwood's plan, I set out to discover the fundamental assumptions that caused modern physicists and cosmologists to cherish such strange ideas. Unfortunately, an extensive search of the literature with regard to the assumptions of science did not result in much. Sure, there was stuff about "nature is orderly" and "causal," and that evidence from nature leads to discoveries of what those causes were, but little that would fit Collingwood's plan. Sure, like Einstein,[67] we

---

[66] Popper, 2002, The Logic of Scientific Discovery.
[67] Einstein, 1905, On the electrodynamics of moving bodies.

frequently mention the assumptions or postulates we are using in our investigations. But most of these are not fundamental. That is, they often do not have opposites and many are falsifiable. The key turned out to be, as is often the case, to look for controversies. The more heated the discussion, the more likely that the opposing viewpoints stem from opposed fundamental assumptions, none of which can be proven correct without a doubt.

As it turns out, my predilection to discover those assumptions stemmed from an assumption too: determinism, the assumption that there are material causes for all effects. But as an idealist, Collingwood obviously was not sure about determinism. He never discovered what the assumptions of science were, and probably was not interested. The same holds for Thomas Kuhn, whose classic, "The Structure of Scientific Revolutions,"[68] showed how a scientific paradigm eventually exerts theoretical control of a discipline. He did imply that important assumptions need reevaluation in the early stages of a paradigm shift, but he never stated what these fundamental assumptions were. He did leave behind one crucial clue to digging them up: his observation that the popularity of a scientific paradigm is dependent as much on the sociological context as on the experimental context.

This meant that I should concentrate, not so much on the details of science, but on the milieu in which it existed. I first came across the debate between determinism and free will as a freshman in college. I tended to support determinism in these debates, even though I was a devout Lutheran at the time. I had taken a slew of freshman science courses at U.W. Madison, and I must have learned something. The liberal arts students tended to support the free will side of these arguments. The battle was endless; there never was a clear winner. This was my second clue

---

[68] Kuhn, 1962, The Structure of Scientific Revolutions.

as to what was important: there was no way anyone could finally prove who was right. There was a perpetual struggle between the two opposing philosophies, determinism and indeterminism. Others, the dialectical materialists, considered the philosophical struggle to be between materialism and idealism. Of course, in science, we used both materialism and idealism a lot. Every model we proposed was necessarily an idealization. That was not a suitable dichotomy; it had to be determinism vs. indeterminism, which was decidedly unpopular among dialectical materialists[69] who evidently thought that their hoped-for revolution would be impossible without free will.

This meant that if there were material causes for all effects, then everything that happens was natural. One could not make a single decision without it being the result of a long chain of causal events. I adopted this radical stance. This became my cardinal rule: If any particular interpretation of scientific data resulted in a free will conclusion, I discarded it. The next step was to elaborate on this realization. One could hold one assumption of science, or hundreds of them. I choose ten for pedagogical and traditional reasons. Each of them supports the scientific worldview by emphasizing points of contention in a slightly different way. You need to understand the "Ten Assumptions of Science," because otherwise I do not think you will understand Infinite Universe Theory.

First: Materialism

The external world exists after the observer does not.

Stated this way, the most important assumption of science, materialism, seems obvious and only common sense. Although the ultimate test cannot be performed personally, we see plenty

---

[69] Engels, 1883, Dialectics of nature, p. 218.

of evidence supporting the assumption. **Materialism**[70] is what gives science its notorious penchant for "objectivity." We assume that the real world consists of objects and that these objects are made of matter. The opposite assumption, *immaterialism*,[71] assumes that the universe is an illusion or internal perception that would not exist without a perceiving being.

Humans often behave in ways that reflect immaterial beliefs. For example, infants sometimes cover their heads with blankets to make the rest of the world disappear. Adults sometimes doubt that a tree falling in the woods does make a sound. Modern physicists follow Einstein in claiming that gravity is not caused by objects, but by "curved empty space" or an immaterial field. Quantum physicists, especially, are not sure about the existence of objects not perceived. Nonetheless, the widespread belief in **materialism** has led to the formal use of observation and experiment as the arbitrators of truth, not only in science, but also in the legal system, as well as in secular society generally. The analysis in this book assumes the universe operates materially, regardless of our ability to sense or detect many of its infinite operations. The most important observation with regard to **materialism** was made by Isaac Newton, the greatest scientist[72] who ever lived. Beset by *immaterialism* on all sides, Newton courageously proclaimed his laws of motion, which operated naturally without mention of divine intervention.

Second: Causality

All effects have an infinite number of material causes.

---

[70] Deterministic assumptions are in bold italics.
[71] Indeterministic assumptions are in italics.
[72] Actually, Newton (1642-1726) was a "natural philosopher." The term "scientist" was not used until 1834. See: Ross, 1962, Scientist: The story of a word.

Although *materialism* assumes existence and location for each xyz portion of the universe, it states nothing about what those portions of the universe do. The object in motion or at rest hypothesized in Newton's First Law remains static, unchanged forever without the action of his Second Law, which describes causality. A cause occurs when one object collides with another, an interaction described by the equation, F=ma, with force, F, mass, m, and acceleration, a. We define the faster object as the collider and the slower object as the collidee. In other words, the impact of the collider causes an increase in the velocity of the collidee. The effect changes the velocity and position of the collidee as well as that of the collider. In addition to *materialism*, causality serves as the basis for modern science. It is the fundamental assumption of the philosophy of determinism, while its indeterministic opposite, *acausality*, assumes that at least some portions of the universe are not subject to material collisions from other portions of the universe. Except in modern physics, scientists normally reject assertions that material collisions are not necessary to produce effects. That is why scientists do not believe in ghosts, ESP, and a multitude of other paranormal claims.

There are two main types of causality: 1) specific causality and 2) universal causality. Specific causality only pertains to restricted portions of the universe. Because it is restricted, it is consupponible with specific acausality, which assumes that other restricted portions may not be causal. Specific causality allows one to pursue science at work, while holding fast to traditional beliefs in the supernatural at home. Universal causality is the generalization applied to all events. It can be broken down into two types: 1) finite and 2) infinite.

Finite universal causality is the second most important assumption in classical mechanics, in which the number of causes required to produce a particular effect is finite. Despite the presumed revolution instigated by the Heisenberg Uncertainty Principle and by relativity, finite universal causality remains the conventional view.

Infinite universal causality (hereafter referred to as ***causality***) assumes, with David Bohm,[73] that the number of causes for a particular effect is infinite. In essence, this conclusion stems from the assumed infinitely subdividable nature of matter (***infinity***). Instead of Newton's hypothesized empty space, both the collider and the collidee exist in an infinite sea of particles, each of which is in motion. The upshot: no mathematical equation, which necessarily must be finite, could produce perfect predictions. One may presume that, once a collision has occurred, there were a finite number of colliders involved. But because of infinite subdividability, we would never be able to know which of those tiny particles was responsible for the plus or minus variations we observe in every experiment at every scale.

Third: Uncertainty

It is impossible to know everything about anything, but it is always possible to know more about anything.

Because each effect appears to be the result of an infinite number of causes, there is an obvious restriction on what we can know about anything. The practical result is that it is impossible to obtain identical experimental results two times in a row. Again, the infinite divisibility of matter produces variations that are responsible for the plus or minus errors in our results no matter the scale or how precise the determination. This means additional, less significant, causes for an effect might be discovered by improving experimental control. Nonetheless, we

---

[73] Bohm, 1957, Causality and Chance in Modern Physics.

can never discover them all, because the causes appear as infinite and the experimental control can never be perfect.

Uncertainty is often associated with the concept of randomness or chance. There are three ways of viewing chance: (1) as a sign of *acausality*, (2) as a singular cause, or (3) as a sign of observer ignorance. In tune with **causality**, **uncertainty**, of course, must side with observer ignorance. What we typically call chance is simply an effect produced by what appears to be an infinite number of unknown variables. Games of chance do not involve *acausality* or an imaginary factor called chance, as claimed by believers in *finity*. Instead, we assume that every roll of the dice produces collisions that may not be completely predictable, but are causal nonetheless.

Fourth: Inseparability

Just as there is no motion without matter, so there is no matter without motion.[74]

Inseparability is especially difficult for many to understand. On first thought, it seems obvious that something must exist for motion to occur. On second thought, it is not that simple. In the first place, there is no connection between matter and motion, for "connection" only applies to things, not to motions. An object is not part matter and part motion. In addition, we are not assuming that one kind of matter is inseparable from another kind. Ironically, one cannot understand the ***inseparability*** of matter and motion without separating them conceptually. You can get a better feel for this by studying the opposing assumption, *separability*, which is prone to four specific logical errors that violate *inseparability*:

---

[74] Hegel's famous dictum, as presented in Houlgate, 1998, The Hegel reader, p. 270.

Errors

1. That matter could exist without motion
2. That motion could occur without matter
3. That matter *is* motion
4. That motion *is* matter.

Examples of each of these four errors follow.

The first error essentially denies evolution. It was associated with pre-Copernican cosmology wherein Earth was considered as a permanently fixed object—just as created. The unchanging Earth was stationary with the sun, planets, and stars rotating around it. As recently as 1962, when I started out in science, Earth's crust generally was considered motionless too, a notion dispelled thereafter by the theory of plate tectonics. In Chapter 16.1 I will elaborate why neither baryonic (ordinary) nor aether matter cannot exist without being in motion.

The second error reigns today when modern physicists, following Einstein, assert that immaterial, massless fields are the cause of gravity and magnetism. Theoretically, that is not possible, because causes result from collisions per Newton's Second Law of Motion as assumed by *causality*. Unfortunately, this petard commonly befalls those who seek to "reform" relativity. In Chapters 16.5 and 16.6 I show that charge and magnetism require interactions with a macrocosm containing aether particles.

The third error occurs primarily as the common belief that matter (viewed as mass) can be converted into motion (viewed as energy). What actually happens during reactions described by the $E=mc^2$ equation is the conversion of one type of the motion of matter into another type of the motion of matter. I explained this reaction in detail in one of my more popular papers available as a free download on the Internet.[75]

The fourth error is the objectification of motion. The most prominent examples are in Special Relativity Theory, where Einstein objectified time as length and light as matter, and in General Relativity Theory, where he objectified time as a dimension.[76] Time is motion. Because time is not an object, it has no dimensions, and therefore cannot dilate as objects do. This is quite simple for determinists, but it is nearly impossible for indeterminists to understand.

Fifth: Conservation

Matter and the motion of matter can be neither created nor destroyed.

The assumption of *conservation* builds upon the assumption of *inseparability* between matter and motion. The idea behind this assumption began with the conservation of matter only. This was not adequate because it eventually became clear that transformations involving matter also involved exchanges of the motion of matter. Newton recognized this early on in his formulation of the Third Law of Motion. He pointed out that when the momentum of a collidee increased, the momentum of the collider decreased by the same amount. Modern physicists usually state this as the First Law of Thermodynamics, the conservation of energy. Problem is, energy neither exists nor occurs—it is a calculation. As I will explain in detail later, energy, like momentum and force, is a matter-motion term, a calculation in which we multiply a term for matter (which *does* exist) and a term for motion (which *does* occur). The

---

[75] Borchardt, 2009, The physical meaning of $E=mc^2$.
[76] Borchardt, 2011, Einstein's most important philosophical error.

conservation of energy is the conservation of nothing. At best, it would be the conservation of an equation as Feynman implied.[77] That is why I write the Fifth Assumption of Science in terms of matter and the motion of matter. "Energy," like the other matter-motion terms will continue to be convenient shorthand for most folks, particularly those trained in modern physics. Nonetheless, I try to avoid matter-motion terms as much as possible, just as we did so successfully in "Universal Cycle Theory."

Sixth: Complementarity

All bodies are subject to divergence and convergence from other bodies.

Complementarity is an unavoidable consequence of the assumptions of ***inseparability***, ***conservation***, and ***infinity***. All interactions produce divergence in one area of the universe at the expense of convergence in another area. This yin and yang always occurs because there is no such thing as a closed system in the Infinite Universe. The boundaries of every portion of the universe are permeable to the entrance and exit of matter and the motion of matter. In the "neomechanical" model used in this book, I refer to a system as a microcosm and the surrounding region as its macrocosm. Therefore, regardless of the immensity of a microcosm, in an Infinite Universe an infinitely large macrocosm still surrounds it. Because every part of the universe must always interact with some other nearby parts of the universe, the complimentary actions of divergence and convergence occur endlessly at every scale. Hydrogen atoms converge to form stars in which the hydrogen converges to form helium, with the resulting excess motion diverging as sunlight. The light converges on plants, allowing them to grow for a

---

[77] Feynman and others, 1964, The Feynman lectures on physics. See the discussion of Chapter 10 below.

while. Both the star and the plants eventually will die, with their constituents and motions diverging into the environment.

Sharp-eyed readers will note a similarity with Newton's First Law of Motion. Just add *infinity* and Newton's object in motion is bound to hit something. Still sharper readers will note a similarity between the First Law of Motion and the Second Law of Thermodynamics, which states that an isolated system can only become more disordered. If we should draw a fixed sphere around Newton's object in motion, it will inevitably diverge from that sphere. In other words, disorder, entropy, and divergence bespeak of the same phenomenon. Similarly, an increase in order requires convergence per *complementarity*. An isolated system can only fall apart, just as your house would fall apart if left unattended. Cosmogonists who believe the universe is expanding sometimes use the Second Law of Thermodynamics to predict its eventual "heat death." Once you add *infinity*, however, that cannot happen. In an Infinite Universe, the various diverging parts and diverging motions of the so-called "isolated system" of the Second Law of Thermodynamics must converge upon yet another "isolated system." In other words, it is a law describing divergence and its complement is a law describing convergence. That is why I call it *complementarity*.

Seventh: Irreversibility

All processes are irreversible.

Irreversibility deals with the abstraction of motion that we call time. In its broadest application, universal time is the motion of all things with respect to all other things. In its narrowest application, specific time is the motion of one thing with respect to another thing. Again, time is motion, and therefore does not exist—it occurs. Time is not *part* of the universe. It is what its various parts *do*. Time is irreversible because each motion of each microcosm in the Infinite Universe is unique. Folks who still believe that travel into the past might be possible are either delusional Sci-fi fans or victims of relativity.

One way to view it is this:

1. It is a fact that the planets, stars, galaxies, etc. are in motion with respect to each other.

2. That makes the night sky unique. It is never the same even two seconds in a row.

3. "Going back in time" would entail moving those heavenly bodies back to the positions they had on the night targeted for this fanciful adventure. Good luck with that.

The opposing assumption, *reversibility*, underpins systems philosophy, which tends to overemphasize the system and neglect the environment. Lab technicians often believe they can demonstrate *reversibility* by providing a semblance of former experimental conditions. When we ignore the environment, reactions in such systems seem like they are reversible. However, when the environment is included, then each reaction properly appears unique and unprecedented. With perfectly empty space being impossible and with the ubiquity of aether, our inability to produce perfect isolation prevents us from getting exactly the same result each time we perform an experiment. Even though the idea of reversible time makes great stories for science fiction, it holds no relevance in the real world. Prospective time-travelers are destined to be forever disappointed.

Eighth: Infinity

The universe is infinite in both the microcosmic and the macrocosmic directions.

In a sense, without *infinity*, there could be no "Ten Assumptions of Science." It is what makes them consupponible—and diametrically opposed to the Big Bang Theory. The concept of an Infinite Universe is simple. The universe contains infinitely small microcosms of matter in motion and infinitely large microcosms of matter in motion. This assumption has served us well throughout the history of science even though it always faced varying degrees of indeterministic objection along the way. Atomic particles are subdividable with no end in sight; celestial bodies are integrable with no end in sight. After every experimental division or observational integration, the evidence for the opposing assumption, *finity* has taken a beating.

In the past, various thinkers proposed two types of infinity: microcosmic and macrocosmic. For instance, Aristotle considered matter infinitely subdividable, but not infinitely extendable; Newton considered matter finite, but infinitely extendable. Even though, logically, it is only a matter of scale, it is rare to find anyone who combines the two assumptions. Apparently, that is because it is so difficult for many to make the jump from *finity* to *infinity* in one fell swoop. One reader told me that he was 95% in agreement with the Ten Assumptions, with the microcosmic part of *infinity* being the only problem. Of course, that is like being 95% pregnant. In their reluctance to make the switch, agnostics handle the cognitive dissonance by overlooking the fact that it is a logical either/or situation. The universe has to be either finite or infinite and particles must be either finite and undividable or infinite and subdividable. The choice one makes between the assumptions of *infinity* and *finity* determines not only the beginning point, but also the ending point of one's cosmology. Because the universe is infinite, logical arguments about it must be circular: start with ***infinity***, end up with ***infinity***; start with *finity*, end up with *finity*.

Ninth: Relativism

All things have characteristics that make them similar to all other things as well as characteristics that make them dissimilar to all other things.

Relativism contrasts with the alternative of *absolutism*. Absolutists believe that things may be identical or not identical. This may be fine in mathematics, but it is not possible in an Infinite Universe where matter is infinitely subdividable and constantly in motion. No two things are identical and nothing can be the same as it was a moment earlier. Because of this reality, the similarity-dissimilarity continuum implied above is a great tool for gaining knowledge about the universe.[78] As we demonstrated in "Universal Cycle Theory," the assumption of ***relativism*** implies that even unobservable portions of the universe have some similarities with the observable portions.

Tenth: Interconnection

All things are interconnected; that is, between any two objects exist other objects that transmit matter and motion.

Interconnection implies that the universe has both a discontinuous and continuous nature. The universe nowhere contains either empty, discontinuous space or solid, continuous matter. The ideas of absolute discontinuity and absolute

---

[78] Borchardt, 1974, The SIMAN coefficient for similarity analysis.

continuity are only that: ideas. As usual, the reality is in between, a reality we nevertheless cannot express without those ideas. That is why loss-less photon travel through empty space is as impossible as the warping of said empty space. Note that **conservation** and **infinity** imply **interconnection**, and *vice versa*. All portions of the Infinite Universe are subdividable and all things transform into other things only as they gain or lose matter or motion to other things. The opposing assumption, *disconnection*, negates the other assumptions of science as well as the consupponibility rule—by contradicting them. *Disconnection* tolerates the theoretical contradictions and paradoxes so common among indeterministic thought, while **interconnection** does not.

This concludes the brief summary of The Ten Assumptions of Science. If you find any of these difficult, as most do, feel free to read the expanded versions in "The Ten Assumptions of Science" and "The Scientific Worldview." Per Collingwood's criteria for exposing fundamental assumptions, none of these assumptions is completely provable. Even so, they form a nice logical package that appeals to folks who detest contradiction and believe that there are material causes for all effects. On top of that, they are supremely useful in making sense of the Infinite Universe.

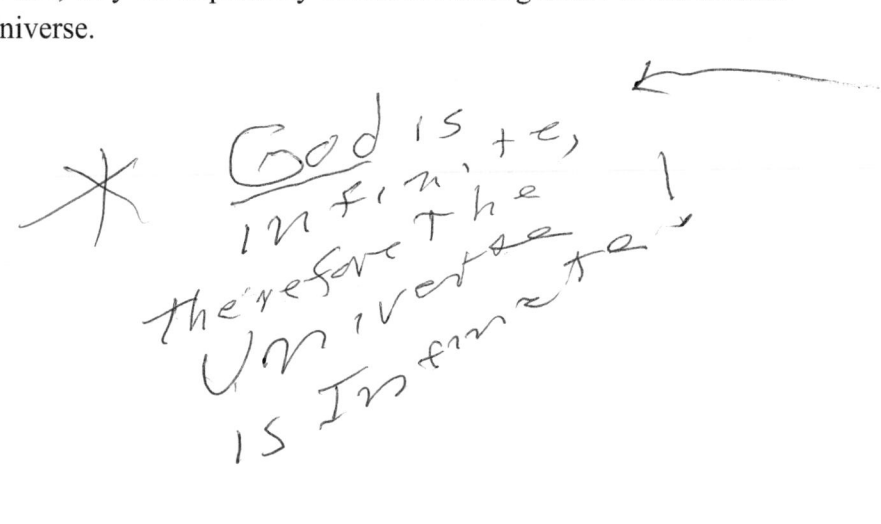

God is infinite, therefore the Universe is Infinite

Chapter 10
Progressive Physics

What has been going on in physics since the beginning of the 20th century is what I call "regressive physics." Without it, the Big Bang Theory would not have survived. It is part of the determinism-indeterminism philosophical struggle, which I summarized in "The Scientific Worldview."[79] Like political evolution, which swings from right to left in a cyclic spiral, philosophy swings from indeterminism to determinism in a cyclic spiral as humanity learns more about the universe. Moreover, like all evolution, this involves the destruction of the unfittest—the abandonment of ideas no longer suited to current, ever-changing conditions. Unlike most other disciplines, theoretical physics is heavily dependent on philosophy. As we saw, interpretations based on indeterministic assumptions can get surprisingly irrational. Other, more practical disciplines tend to be held more closely in check by observation, experiment, and falsifiable assumptions. In addition, when studying a small portion of the universe, one often can get away with assuming *finity* without committing a fatal theoretical error.

As mentioned, this is not so with cosmology because there is no possible way to prove whether the universe is infinite or finite. Nonetheless, the choice between the assumptions of **infinity** and *finity* determines the course of theoretical explanation. As we saw, that choice affects more than just cosmology. It pervades all of physics as well, producing the paradoxes and contradictions disregarded or even celebrated by indeterminists accustomed to the plentiful contradictions in their religious lives.

---

[79] Borchardt, 2007, The Scientific Worldview, pp. 14-23.

There is a way out of this theoretical mess: progressive physics. To do so, we must adhere closely to a set of principles that will provide a firm foundation for proceeding. "The Ten Assumptions of Science" provide that foundation. If those are to be believed, then we need to do a little house cleaning with regard to a few elementary concepts whose misinterpretations lead down the regressive path. The most misleading involve indeterministic interpretations of the common matter-motion terms that we find so useful in physics. As we assume throughout "The Ten Assumptions of Science," the universe presents us with only two basic phenomena: 1) matter and 2) motion. With *inseparability,* this idea becomes a bit more overt: Just as there is no motion without matter, so there is no matter without motion. This deterministic assumption appears particularly difficult for regressives to understand, especially since Einstein's aether denial and his introduction of the concept of "immaterial fields." Although matter and motion are simple concepts, further explanation is needed because they are a major part of the philosophical struggle.

Matter

Matter is an abstraction for all things in existence. As an abstraction, matter *per se* does not exist in the same way that the abstraction "fruit" does not exist. We only get to eat specific examples of fruit, an apple or orange perhaps, but we can never eat a fruit. Similarly, each thing in the matter category is unique per *relativism*. Each thing contains other things inside and is immersed in other things *ad infinitum*. Each thing in existence has three dimensions and location with respect to other things. The three dimensions are given by the three coordinates, x, y, and z. The "location with respect to other things" is specified by comparing at least two sets of coordinates, with the distance between them being a comparison with the distance between any two other objects in the universe. Because all portions of the

Infinite Universe are in continual motion, there is no absolute location or distance to which we can refer. In each case, we must arbitrarily choose a point of reference or "reference frame" for relating one thing or one measurement to another.

Infinity assumes that matter is infinitely subdividable and integrable. Above all, matter is not the "solid stuffing" that we thought in days past. Perfectly solid matter and perfectly empty space are simply ideas. As with all idealizations, they do not and cannot exist despite the fact that we need them for understanding the intervening reality. This means that there can be no ultimate particle filled with perfectly solid matter, just as there can be no portion of the universe that is perfectly empty space. In addition, when we subdivide a thing, that is, a microcosm, we always end up with what we think of as matter and what we think of as space. No matter and no space are devoid of either property. Thus, any subdivision of the universe will produce at least one submicrocosm (what we might consider the "matter" part) that has greater mass and/or greater motion relative to another submicrocosm (what we might consider the "space" part). The subdivision of space is as important as the subdivision of matter. Indeterminists often object to infinite subdividability, saying that it ultimately would imply a solid universe. However, one could just as easily say that it ultimately would imply an empty universe. There is no objective reason to favor one property over the other, because those two properties operate at all scales. Again, perhaps the most important property of matter is mass, which is the resistance to acceleration. Matter and mass are not identical; as I will explain in the neomechanical explanation of the absorption and emission of motion wherein mass increases and decreases, while matter does not.

Any other "definition" of matter will be confusing, as it probably was in your physics class. Per *infinity*, however, its definition must involve a "passing of the buck." Because the universe is infinite, nature can provide no black/white answer for the question: "What is matter?" As much as this bedevils absolutists who cannot understand abstraction and are prone to seek *certainty*, it is not possible to define in finite terms that which is infinite. The best we can do for them is the definition we use in neomechanics: matter is any xyz portion of the universe. Another bedevilment for indeterminists involves the question "Where does matter come from?" For Infinite Universe Theory the short answer is, again, a matter of "passing the buck": Matter always comes from other matter (See Chapter 16.4).

Motion

Motion is the abstraction we use for all occurrences. As mentioned, there are only two fundamental phenomena in the universe: matter and the motion of matter. Of course, with our assumption of *inseparability* we "tie" these two phenomena together: "Just as there is no motion without matter, so there is no matter without motion." Note that I put the word "tie" in quotes. This is because it actually is not possible to tie matter and motion together. There can be no physical "connection" between matter and motion. As neomechanists, we often use "motion" as shorthand for what really is the "motion of matter." Unlike matter, motion does not exist—it does not have xyz dimensions. Motion is what things do. Although motion occurs within the universe, it is not *part* of the universe, for all parts have three dimensions and motion does not. Regressive physicists commonly violate *inseparability* by claiming that motion can occur without matter, as Einstein did in his speculations about "immaterial fields."

Another word for motion is time. Universal time is the motion of all things with respect to all other things and is immeasurable. Clocks measure specific time: the motion of a particular thing with respect to another. We use those measurements to represent time, but time is not measurement, as is so often claimed by indeterminists. The dinosaurs experienced time (motion), but they did not measure it.

Regrettably, many folks are like Newton, who violated ***inseparability*** by imagining that "absolute time" occurs independently of matter: "Absolute, true and mathematical time, of itself, and from its own nature flows equably without regard to anything external..."[80] Because time is motion, it cannot flow. Flowing is motion. Saying that time flows would be like saying that motion moves, which makes no sense. For there to be flowing, there has to be something flowing; for there to be motion, there has to be something moving. That is quite simple, but it appears difficult for most folks to accept that time is motion. We are struggling to escape from the conundrums left behind by Newton and Einstein. One important characteristic of regressive physicists is the inability to know what time is. The average person seems to think that time is a great mystery or that "it" flows or that one could go back in time, as if it was a thing like a house with receding doorways. Again, many with solipsistic tendencies, especially those in quantum mechanics, believe that time does not occur unless it is observed or measured. That is our background, and it takes each of us a while to overcome the propaganda surrounding such a simple phenomenon. The claims that "time is a measurement" or that "time is an illusion" characteristically strive to keep the solipsistic observer in the picture.

The realization that "time is motion" is so radical that halfway measures sometimes seem preferable. A common one is the belief that time is an "aspect of motion," as if motion had other aspects independent of time. Time does not exist, it occurs. Only things, having xyz dimensions, can exist and have properties and

---

[80] Newton, 1687, Principia, p. 77.

aspects. Time does not have properties or aspects. The aspect idea therefore is a violation of ***inseparability***. Another violation involves the idea that, in order to be considered "time," a particular motion must be cyclic. This is indeed strange, because that would mean that the linear motions envisioned in Newton's laws of motion would take place in no time. Object A could collide with object B without taking any time to do that. I once complained about a suggestion that catastrophe theory could describe events occurring in zero time.[81] The complete liberation only comes when we finally realize that "time is motion." And as maintained by Lucretius over two thousand years ago, "time by itself does not exist…It must not be claimed that anyone can sense time by itself apart from the movement of things…"[82]

Matter-Motion Terms

To help understand these two primary phenomena, physicists coined many commonly used "matter-motion terms."[83] Matter-motion terms represent calculations that generally involve the multiplication of a term for matter and a term for motion. Thus, momentum is $P=mv$, where m represents mass (matter) and v represents velocity (motion). Other common matter-motion terms are force ($F=ma$) and energy ($E=mc^2$). Unfortunately, by combining a term for matter and a term for motion, we create a creature that neither exists nor moves. We cannot take a piece of momentum, force, or energy home with us and momentum, force, and energy cannot exhibit motion. These are just calculations that help us understand our measurements of matter in motion. They are not actual things.

---

[81] Borchardt, 1978, Catastrophe theory.
[82] Lucretius, 60 BCE, On the nature of the universe, p. 464.
[83] I should trademark the phrases "regressive physics" and "matter-motion term," as no one else seems to use those combinations with the definitions I use for them.

A special problem arises when we use matter-motion terms tens of thousands of times, as physicists do. We tend to forget that matter-motion terms are only calculations. Extensive use of the shorthand for these terms when we do the math only makes it worse. I would bet that most physicists believe that momentum, force, and energy exist as physical objects or motions. Most folks talk about "energy" as if it was something that they could take home with them to save for a rainy day. Some appear to believe that, upon ignition, energy might come flowing out of a burning stick or lump of coal. But, as mentioned, energy does not exist; it is a calculation. That is what I learned before I could write one of my most important papers[84]: "The physical meaning of $E=mc^2$." I did not realize that I had to step over an ever-widening philosophical divide just to figure out what a simple equation meant. Later in the text, I will present more details concerning this divide. If you can follow the above paragraphs, then you are beginning to stare into the abyss that must be crossed before you can understand what will replace the Big Bang Theory.

If you ask a regressive physicist or cosmogonist the meaning of any of those terms, you will not get a straight answer. About the closest you will get is this by Prof. Richard Feynman discussing that the Conservation of Energy was just the conservation of a calculation, not of anything or of any motion:

> There is a fact, or if you wish, a law, governing natural phenomena that are known to date. There is no known exception to this law; it is exact, so far we know. The law is called conservation of energy; it states that there is a certain quantity, which we call energy, that does not change in manifold changes which nature undergoes. That is a most abstract idea, because it is a mathematical principle; it says that there is a numerical quantity, which does not change when something happens. It is not a description of a mechanism, or anything concrete; it is just a strange fact

---

[84] Borchardt, 2009, The physical meaning of $E=mc^2$.

that we can calculate some number, and when we finish watching nature go through her tricks and calculate the number again, it is the same.[85]

Because energy is not matter, motion, or a physical combination of the two, its calculation lends itself to numerous indeterministic interpretations. Within regressive physics, one of the most popular myths is that mass can be converted into energy.[86] Another is the claim that forces exist. Forces neither exist nor occur. Like energy and momentum, force is an extremely valuable calculation. In the equation F=ma, what *does* exist is the matter contained in the xyz portion of the universe that has the mass, m. What *does* occur is the change in motion represented by the letter a, which represents the acceleration produced when that matter collides with something. There are no forces that exist or occur independently of matter. When your sailboat is pushed by the wind, you are being pushed by the wind, not by some independent "force" floating around in the air. We can calculate that effect with our valuable equation F=ma, but we cannot take that "force" home with us despite Darth Vader's valediction.

Similarly, momentum, P=mv, describes a body with mass m with a velocity v. There is no such thing as momentum. There are things having resistance to acceleration that we measure as mass, m, that may undergo motion we can measure as velocity, v, but nothing contains or exhibits momentum. Instead, things contain matter and exhibit motion. For instance, when you are hit by an offensive tackle having mass m and velocity v, you still have been hit by an offensive tackle, not by momentum or by a calculation or concept called momentum. When the tackle is small and slow, we calculate that the momentum will be less than when the tackle is large and fast. The calculation helps us

---

[85] Feynman and others, 1964, The Feynman lectures on physics.
[86] Borchardt, 2009, The physical meaning of $E=mc^2$.

understand matter in motion. The calculations present two different pictures of what might happen. We must always remember that matter-motion terms like momentum are like pictures of a running dog; they are not the running dog.

The misuse of matter-motion terms is one of the defining characteristics of regressive physics. That came about for many interlocking reasons. Firstly, it is easy for anyone who uses such shorthand terms all day long to objectify them, giving them a teleological connotation. They become like friends. Secondly, regressive physicists are notorious for being immaterialists. As mentioned, Einstein, for instance, thought of gravitational and magnetic fields as being "immaterial" and quantum mechanists believe that an unobserved particle does not exist. Thirdly, having eschewed mechanism, regressives tend to avoid the proper terms "matter" and "motion." That is why the Fifth Assumption of Science, **conservation** (Matter and the motion of matter can be neither created nor destroyed) is stated by them as the conservation of energy. That misuse of the term appears to have been particularly rampant in the 20th century. That is why they cannot define energy, and that is why some of them think of "dark energy" as being an independent constituent of the universe. That is why some true believers even claim that the whole universe was filled with energy before it was converted into mass ($E=mc^2$ don't you know?).[87] Fourthly, because the universe is microcosmically infinite, we always reach a point when, like Newton and Einstein, we see evidence for motion but cannot see evidence for its material carrier. In other words, we see the effects, but not the microcosm that is producing the effects. Good positivists like Einstein, can calculate a force in any case and not be bothered by not knowing what caused it. The math works whether the things involved are real, imagined, or

---

[87] Note that $E=mc^2$ is a perfectly good equation when interpreted correctly, as explained in Borchardt, 2009, ibid.

nonexistent. Indeterminism does not require the unseen actor to be anything other than their imagined, mysterious, matterless force. None of that prevents experiments from confirming the mathematical suspicions when real microcosms are involved.

On the other hand, you might agree at least partially with ***inseparability***. This seems only commonsense. How could there be motion without something moving? The idea that matter cannot exist without being in motion is less obvious, but just as true as the first half of the assumption. We cannot find any motionless matter—even near absolute zero. Could it be that motion is required for matter to exist? Later, we will explore this in Chapter 16.1.

In the popular press, there is a great fuss about the search for the grand unification of the "four forces," which are considered as if they were actual things instead of the results of things hitting things. The things exist, but the forces do not. Again, like the other matter-motion terms, force is a calculation. When we write: F=ma, we are writing a statement about the microcosm (m=mass) and its motion (a=acceleration). The "force" of gravity is a description of the collision of one type of microcosm with another. Because they are not sure about what is doing the colliding, positivists and operationalists tend to objectify matter-motion terms.[88] In "Universal Cycle Theory," we dispensed with as many matter-motion terms as we could by using matter and motion instead. The upshot: Giving up "force," for example, made us look for that which is doing the "forcing" instead. That neomechanical approach is what made that book so successful.

When one asks if something is real, that means the same as: Does a thing exist? Existence is a property only of microcosms, portions of the universe that have three dimensions and location

---

[88] Borchardt, 2011, Einstein's most important philosophical error.

with respect to other microcosms. Therefore, in neomechanics we say, for instance, that the brain exists but that the mind does not; legs exist, but running does not. Both thinking and running occur. These motions are real phenomena just like the matter that produces them, but unlike matter, they do not have dimensions and do not exist. A dead organism may have a brain, but it will not have a mind. No mind or consciousness can exist independently of its associated brain despite Deepak Chopra's absurdly indeterministic claims.[89]

Our adoption of *infinity* forces us to move beyond traditional views, which were based entirely on *finity*—as demonstrated by the continued existence of cosmogony as a serious subject. Like the Big Bang Theory, both classical mechanics and relativity were founded on the assumption of *finity*. Because the universe is infinite, that assumption was destined to run its course, as seen by the numerous contradictions it engendered. Still, human minds tuned to finite mathematics, generally resist the idea that the universe continues on and on forever. Nonetheless, the use of *infinity* in the study of matter in motion (mechanics) is such a dramatic step I call it "neomechanics." That term is warranted because the simple explanation I present in the next chapter would not be possible without *infinity* and could not originate with regressive physics beset by aether denial. If one assumes that matter is infinitely subdividable, then aether must exist, along with its even smaller constituents. Otherwise, the absorption and emission of motion could not occur in accordance with mechanics and Maxwell's $E=mc^2$ equation of 1862.

---

[89] Chopra and Kafatos, 2017, You are the universe: Discovering Your Cosmic Self and Why It Matters. [Chopra is probably today's foremost immaterialist. Like his predecessor, Bishop Berkeley, he claims that the existence of the universe depends on his consciousness.]

Chapter 11

Neomechanics

In true philosophy one conceives the cause of all natural effects in terms of mechanical motions. -Christiaan Huygens, 17th Century[90]

The Eighth Assumption of Science, **infinity**, changes everything. The Ten Assumptions of Science were consupponible, coherent, and non-contradictory only because they included that assumption. Huygens was speaking for classical mechanics, which was based on *finity* and was not without subtle obeisance to the supernatural. Along with "The Ten Assumptions of Science," what I call "neomechanics" provides a new foundation for scientific philosophy that promises to lead the way out of the Special Relativity Theory-General Relativity Theory-Big Bang Theory mess.

Note that Big Bang theorists do not and cannot dwell on the fundamental assumptions that underlie their theory. Like Halton Arp and other empiricists, they seem to think that they "Don't need no stinking assumptions." Otherwise, they could not get to first base with the Big Bang Theory. If one had a steadfast belief in **conservation** (Matter and the motion of matter can be neither created nor destroyed), then the Big Bang Theory or any other cosmogony would be plainly unimaginable. Note that cosmogonists even disagree on whether the laws of physics existed before the Big Bang. Hawking, for one, writes "Because there is a law such as gravity, the universe can and will create itself from nothing...,"[91] conveniently forgetting about **conservation**. Contradictions like these go unnoticed among regressives, probably because they tend to worship their heroes and have many other contradictions they need to ignore as well.

---

[90] Huygens, 1690, Treatise on light.

[91] Hawking and Mlodinow, 2012, The grand design, p. 180.

Converting Mechanics to Neomechanics

What do we mean when we say there are material causes for all effects? It is simple. We say that the cause of the home run was the collision of the bat against the ball. That is what mechanics is, things colliding with things. Like the bat and ball, all things consist of matter, which are xyz portions of the universe that have location with respect to other things. A "mechanic" makes sure that the various parts of a machine are colliding properly. A "mechanism" is an explanation of how collisions produce a certain effect. A "mechanist" is one who believes, as I do, that the universe consists only of matter in motion.

Isaac Newton, generally considered the founder of classical mechanics, proposed three simple laws explaining how this came about. By studying his First and Second Laws of Motion, you will be well on your way to understanding how the universe works, barring a few changes I will propose.

The First Law is simply the most important observation ever made:

> *Every body perseveres in its state of rest, or of uniform motion in a right line, unless it is compelled to change that state by forces impressed thereon.*[92]

With this law, Newton essentially is observing that portions of the universe display two primary phenomena, matter and its motion. The matter part, the body, moves with respect to what he thought of as absolute space. The First Law describes inertia or momentum. A body undergoing inertia requires no engine and cannot increase or decrease its motion by itself. This type of motion "keeps" the electrons, planets, and satellites in orbit. It is

---

[92] Newton, 1687, Philosophiae Naturalis Principia Mathematica.

responsible for the action of the Second Law of Thermodynamics, the Principle of Least Action, and the Principle of Least Effort. In other words, inertia is what makes the world go round. Everything that happens is the result of changes in inertia.

Although Newton's First Law is highly idealistic and therefore not exactly correct,[93] the basic idea behind it will persevere. However, in light of our assumption of *infinity* and its correlative assumptions, we need to make some significant modifications, which I will now outline in some detail.[94] These changes may have little practical significance, but they are now clearly necessary for understanding Infinite Universe Theory:

We need to replace the word "body" with the word "microcosm." Both terms describe xyz portions of the universe, but "body" is too vague for understanding the universe from the standpoint of *infinity*. A microcosm always contains submicrocosms, each of which contains subsubmicrocosms ad infinitum.

That gives:

Every microcosm perseveres in its state of rest, or of uniform motion in a right line, unless it is compelled to change that state by forces impressed thereon.

---

[93] Idealizations are ideas about how the world works. Because the universe is infinitely complicated, no single idea about it can be complete. Thus, "perfectly empty space" and "perfectly solid matter" are idealizations. Neither of them actually exists, but such ideas help us find doorways and avoid walking into walls. Right off the bat we know there actually is no such thing as a straight line. The rotation of Earth makes every "straight line" into a curve. Even so, Newton's idealization is good enough for many explanations.

[94] These necessary modifications are some of the reasons for the "Beyond Newton" in the subtitle of my most important book, "The Scientific Worldview."

Since Newton, we find no body at rest with respect to anything else. So, we do not need the phrase "its state of rest."

That gives:

Every microcosm perseveres in uniform motion in a right line, unless it is compelled to change that state by forces impressed thereon.

Since Newton, we are unable to find a right line (or straight line) except in our imaginations. Real right lines are properties of real microcosms, which always contain submicrocosms in motion. Thus, the boards you use to build your house may be straight, but they will never be perfectly straight. In addition, temperature changes in the environment cause that lumber to shrink and swell, twist and turn. If we expand our viewpoint, outside Earth, for instance, we find that because Earth is rotating, any two points describing a right line are rotating as well. The inertial motion of satellites in orbit is nicely described by the First Law—except that it is plainly not in a right line. Still, Newton's visualization of a body traveling through empty space far from any external influences is useful—to some extent.

That gives:

Every microcosm perseveres in uniform motion, unless it is compelled to change that state by forces impressed thereon.

In an Infinite Universe, other microcosms eventually will change the state of any particular microcosm. The word "unless" must be replaced by the word "until."

That gives:

Every microcosm perseveres in uniform motion until it is compelled to change that state by forces impressed thereon.

The phrase "compelled to change" hints at teleology in which the body is being forced to change itself—an impossibility.

That gives:

Every microcosm perseveres in uniform motion until the direction and velocity of its motion is changed by forces impressed thereon.

Similarly, the word "perseveres" gives the teleological impression that the body in question must struggle to maintain its motion. A more objective replacement for "perseveres" would be the word "continues."

That gives:

Every microcosm continues in uniform motion until the direction and velocity of its motion is changed by forces impressed thereon.

As mentioned in the chapter on progressive physics, there are no such things as "forces." The direction and velocity of a body is changed by a collision caused by another body. In the equation for force, we multiply the mass of the collider times its effect on the velocity of the collidee.

Finally, that gives:

Every microcosm continues in uniform motion until the direction and velocity of its motion is changed by collisions with supermicrocosms.

The Second Law explains what we mean by causality:

> *The alteration of motion is ever proportional to the motive force impressed; and is made in the direction of the right line in which that force is impressed.*[95]

A cause is related to the alteration of motion: the greater the interaction between microcosm and macrocosm, the greater the effect on the microcosm. Because Newton dealt with imagined

---

[95] Newton, 1687, Philosophiae Naturalis Principia Mathematica, p. 158.

ideal bodies, the forces were ideal, exact, "ever proportional," and the result was motion along a *perfectly* straight line. These idealistic notions are characteristic of the now obsolete belief in finite universal causality, which distinguished the mechanical view of the world, the philosophy of mechanism. To convert to neomechanics, some changes are necessary for the Second Law as well:

In neomechanics, there is no such thing as a "motive force" that could be impressed upon anything. Instead, all we have are things colliding with things. While forces do not exist, we can calculate the alteration of motion by using the force calculation: F=ma, in which the effect is proportional to the mass of the impacting supermicrocosm multiplied by the change in velocity induced in the microcosm.

That gives:

The alteration of motion is ever proportional to collisions from supermicrocosms; and is made in the direction of the right line in which those supermicrocosms were traveling.

Again, in neomechanics, there are no right lines and no effect is produced by a finite number of supermicrocosms.

That gives:

The alteration of motion is ever proportional to collisions from supermicrocosms; and is made in the direction in which those supermicrocosms were traveling.

You will realize the importance of these changes and the concepts behind them in formulating Infinite Universe Theory. There are other changes proposed for classical mechanics. Although seldom mentioned, and as I will explain later, the advent of the Heisenberg Uncertainty Principle signaled the end for classical mechanics and its foundational assumption of *finity*. There were two different directions that theorists could have gone as a result: 1) deterministic neomechanics or 2) indeterministic relativity. Relativity presented the easiest path, fitting nicely with the indeterministic culture of the times and explaining experimental results unexplained by classical mechanics.[96] Gleefully substituting mathematics for mechanics, relativity required no rejection of *finity*, the long-cherished assumption of the old scientific world view (classical mechanics) and the soon-to-be cherished assumption of the new scientific world view (systems philosophy).

But "The Scientific Worldview" taught us to leave behind the worst of both classical mechanics and systems philosophy. We would no longer overemphasize the outsides of things, as in classical mechanics. And we would no longer overemphasize the insides of things as in systems philosophy. To this effect, I coined a new term to describe the resulting universal mechanism of evolution: univironmental determinism (What happens to a portion of the universe is determined by the infinite matter in motion within and without). The stripped-down version, neomechanics, applies to the laws of classical mechanics, but uses as its "body" an infinitely subdividable "microcosm" surrounded by an infinitely subdividable and integrable "macrocosm." Newton and Leibniz came close to envisioning neomechanics and Infinite Universe Theory when they invented calculus (Chapter 13.2). But the overt assumption of *infinity* was too radical for the times, as the immolation of Bruno demonstrated.

Today, in neomechanics, Newton's body of the First Law of Motion no longer travels through absolute space, conceived as being perfectly empty. It is replaced by a microcosm containing

---

[96] One example is the Ives-Stilwell experiment, the results of which were recently explained by Bryant with orders of magnitude better accuracy and precision without invoking time dilation or anything else to do with relativity (Bryant, 2016, Disruptive, pp. 250-256).

an infinite number of smaller "submicrocosms" traveling through a macrocosm containing an infinity of larger as well as smaller "supermicrocosms." Note that neither the "sub" prefix nor the "super" prefix has any special connotation other than one of location: "sub" for being inside, and "super" for being outside the microcosm of concern.

Note again that Newton's concept of "absolute space" is an idealization only suited to mathematics. We are unable to find any perfectly empty space that could satisfy the "absolute space" designation. We gave up the idea that space is some kind of empty container for things. The perpetual motion hypothesized in the First Law cannot occur either. In an Infinite Universe, there is always something in the way that will produce a collision that will slow Newton's body down. Of course, most of the supermicrocosms in space are tiny and usually too insignificant to interfere much with Newton's idealization. That is one reason classical mechanics has been so successful. Again, the observation that the goings on in the natural world are produced by inertial motion will stand forever as the most important "Law of the Universe."

Because inertia is involved in all occurrences throughout the universe, every natural law ever devised is a subset of Newton's First Law of Motion. This is especially obvious when we swap the assumption of *finity* for the assumption of *infinity*. That modification intensifies and extends the First Law, eliminating any space for the supernatural. As in classical mechanism, we assume that phenomena must fit either of two fundamental categories: matter or the motion of matter, the two abstractions mentioned previously. The inclusion of *infinity* just completes the job at every scale, extending mechanics from the infinitely small to the infinitely large.[97] The upshot is that matter in motion

---

[97] Puetz and Borchardt, 2011, Universal Cycle Theory.

exists everywhere and for all time. Perfectly empty space, that is, nonexistence, is impossible in an Infinite Universe. That is why we cannot produce a perfect vacuum or find a part of the universe without temperature, which is due to the vibrations of matter.

Cartoon Version of Neomechanics

This section continues our little journey to set physics straight. The indeterministic counter-revolution set in motion by Einstein must be reversed. We must return to the deterministic basics encouraged by Huygens, who correctly conceived "the cause of all natural effects in terms of mechanical motions."[98] For over two centuries, that concept continued to bedevil religionists and their indeterministic offspring. Finally, Einstein's regressive step away from mechanics provided welcome relief despite the ensuing contradictions. The cartoons and much of the discussions I present below are adapted from Chapter 5 of "The Scientific Worldview." These are rather simple reductions from classical mechanics in light of the assumption of *infinity*. Of course, there is no way to illustrate *infinity*, much less as a cartoon, but that assumption forces us to recognize that:

Microcosms always contain submicrocosms within and supermicrocosms without.

Neomechanics applies to every portion of the universe, which we assume to be infinitely subdividable and infinitely integrable. What we think of as "perfectly empty space" merely contains matter that poses reduced resistance to whatever matter we do not consider "perfectly empty space."

If anything, the illustrations to follow show that you can gain a rudimentary understanding of the universe simply by studying

---

[98] Huygens, 1690, Treatise on Light.

its two most fundamental phenomena: matter and the motion of matter. Realize again that the terms "matter" and "motion" are abstractions. Matter *per se* does not exist and motion *per se* does not occur. The universe only presents specific examples of matter and specific examples of motion. Per **relativism**, no two such examples are identical. Still, we love our abstractions, generalizations, and stereotypes because they help us discuss, communicate, and understand specific portions of the universe and their motions. Nonetheless, we always must remember that they are mental constructs. Like all idealizations, they do not and cannot exist.

This is the beginning point from which we will attempt to understand the interactions of microcosms and macrocosms. Although all interactions are infinitely complex, they have at the same time a fundamental character to which they may be reduced for elementary discussion. This chapter is extremely important for understanding how the universe operates. In these pages, for example, you will find the scientific foundation for rejecting hypotheses involving extrasensory perception and other claims of the paranormal. I promise that, after you understand neomechanics, you will have a better understanding of $E=mc^2$ than Einstein ever did.

For the determinist, all microcosms are matter and all interactions involve the motions of matter. However, we must be aware that such a reduction is always considered by indeterminists as extreme in the worst sense and is generally resisted with all the means at their disposal. Indeterminists have another goal, one that is equally extreme: the reduction of all things to spirit or matterless motion.[99] The intermixing of determinism and indeterminism usually assures that neither of

---

[99] Some regressive theories in physics claim things consist only of waves or that motion can create matter out of nothing.

these extremes is taken seriously. The indeterministic agenda is for the most part hidden, but because it is always difficult to confront the material world with a spiritualistic reduction, it commonly takes the form of the argument against reduction itself.

Of course, if things are infinitely complex, then to know anything, we must reduce the complex to the simple—that is what Newton did and that is what the illustrations below do in spades. Again, the main difference between classical mechanics and neomechanics is our assumption of *infinity*. Both the classical mechanists and their antireductionist critics were misguided by assuming *finity*. They may be excused for believing that complete descriptions are possible, but with neomechanics, we have no such delusion. That admission is drastic, for taken to extreme, it precludes the possibility of a finite particle and finite universe. Those concepts are dear to the hearts of today's regressive physicists and cosmogonists, making their rejection decidedly new. The "neo," meaning "new," in neomechanics is definitely warranted.

Now, this chapter presents an explanation so elementary that it is akin to the alphabet. Why is this necessary when anyone schooled enough to read this book surely must know the scientific alphabet already? It is necessary because even advanced "scientific" reasoning tends to use a mishmash of determinism and indeterminism—an alphabet derived from two different languages. A few foreign characters may be tolerable, but when there are too many, they produce grotesque theories and interpretations that fail. We have to go back to the beginning and start over again. Indeterminists resist this process, partly because of what they have invested in developing the compromises involved in current explanations, and partly because of what they find whenever they explore the character of their starting points in relation to the character of the universe.

Although the antagonism between thought and action contributes to it, much of the anti-reduction movement is a manifestation of the indeterminist's opposition to the reduction of the world to matter instead of spirit. For many, it is frightful to give up the spirit in the sky and the promise of an eternal afterlife. That is why it is part of the philosophical struggle.

Selecting the Microcosm of Concern

To do justice to Newton's mechanics, we need to mention Newton's Third Law of Motion:

> *To every action there is always opposed an equal reaction: or the mutual actions of two bodies upon each other are always equal, and directed to contrary parts.*[100]

In other words, not only is the motion of the microcosm changed because of its interaction with the macrocosm, but also so are the motions within the macrocosm, according to the Third Law. If one body increases the motion of another, then its own motion decreases in exactly the same amount, in complete agreement with ***conservation***. In the Newtonian view, these changes occur to each body as a whole and do not require the participation of parts or of submicrocosms within. Following in the atomistic vein, Newton's idealization does not require us to assume anything about the internal characteristics of each body. For his purposes, the constituents of each body could be identical and insignificant. It is as if each body was filled with solid matter or empty space. Of course, since then we have learned that there is no such thing as "perfectly solid matter" or "perfectly empty space." Every portion of the universe always contains other things, as will be illustrated below.

Nonetheless, in its heyday classical mechanics was an indispensable argument in the conflict with indeterminism. Its

---

[100] Newton, 1687, Principia.

development accompanied and advanced early versions of many of the Assumptions of Science, and greatly aided the belief that the universe was orderly and understandable. It initiated the view that motion as well as matter was equally important and predictable. In spite of this, the primitive, now indeterministic idealizations of classical mechanics brought about its downfall as the basis for an adequate natural philosophy.

Today, to be known as a "mechanist" is to be linked with these discredited aspects of the Newtonian program. Still, if one accepts *inseparability*, then one is a mechanist of some sort. Any new version of mechanics, however, must be consistent with the scientific assumptions made unavoidable and called for at this time. In particular, it must eschew the microscopic finity and macrocosmic bias of the classical construction. In the proposal below, which I call *neomechanics*, I attempt to weave the Ten Assumptions of Science into a presentation that shows what modifications are necessary in our most elementary picture of the things and events of the world.

Six Interactions between Microcosm and Macrocosm

The univironmental relationship between a microcosm and its macrocosm can be expressed a little more concretely as an interaction between submicrocosms (portions of the universe *within* the microcosm) and *supermicrocosms* (portions of the universe *outside* the microcosm). We begin to understand the nature of this by reducing the infinite variety of real interactions to one idealized interaction: that of matter with matter. For the Newtonian mechanist, this interaction involved portions of the universe that were filled with "solid matter." For us, it involves portions of the universe that simply contain other portions of the universe. This is a significant step beyond classical mechanics and deserves at least a modest explanation.

The reduction discussed below is important for understanding Infinite Universe Theory for the same reason that the Newtonian reduction was necessary for understanding the mechanical worldview. In the most abstract way, it gives us experience in thinking about things and their motions as the real producers of phenomena. In accord with ***inseparability*** we offer no explanation that entertains the ideas of motion without matter, matter without motion, or, in this age of Einstein, matter *as* motion or motion *as* matter. This way of thinking is extremely powerful because, with one stroke we can eliminate the possibility of such "things" as ghosts and spirits, which, it is claimed, do not contain matter and do not act on physical contact.

Note that Newton's reduction, however, was as quintessentially macrocosmic as the reaction to it (systems philosophy) was microcosmic. Remember that macrocosmic thinking tends to overemphasize the outsides of things and deemphasize the insides of things. This was simply adopted from atomism. Newton's laws were simple idealistic reductions that worked well with "objects" or "bodies" assumed to have these characteristics:

They were perfectly rigid.

They were perfectly solid.

They were perfectly inert.

They were perfectly spherical.

These assumptions were nothing new and were not actively defended by Newton. Even though most bodies obviously did not have the above properties, it was widely thought that the fundamental constituents of matter did. Newton's major contribution was to apply mathematics to that vision. By taking his cue from the atomists and their notion of the ultimate, finite particle,

Newton ignored the insides of his model. He ended up treating only one ideal end member of the continuum. By ignoring and thus ultimately rejecting microcosmic infinity, he could simplify objects, treat them as wholes and ignore the complicating aspects of the interactions of their parts. Moreover, even when he found it necessary to invent the calculus, the formal recognition of parts and wholes in mathematics, he stuck to the view that the whole is equal to the sum of its parts. This was the primary failing of Newtonian mechanics. It was a reduction well suited to the mathematical approach, but in hindsight, woefully inadequate for describing the real world. The real world was not made up of identical atoms. If that were true, nothing new would ever come of the interactions of those identical atoms containing solid matter. They would bounce around forever, having no reason to combine to form the myriad of things we observe everywhere. Neomechanics uses the opposing assumptions, **relativism** and **infinity**, while Newton's idealization uses *absolutism* and *finity*. This was despite his co-invention of calculus, which deals with the infinitely small and infinitely large, portending Infinite Universe Theory (Chapter 13.2).

Despite ignoring the constituents of his object of concern, Newton's belief in *finity* resulted in a microcosmic error as well. Microcosmic errors overemphasize the microcosm and neglect the macrocosm. Ernst Mach (1838–1916), a prominent Austrian physicist and positivist philosopher, rightly criticized Newton's First Law by saying that nothing can be significantly predicated of a body's motion if the rest of the universe is assumed to vanish. Newton, of course, thought of absolute space as a point of reference when he idealized the motion of the body in the first part of his proposition. Some physicists have vainly attempted to discover a fixed point of reference, such as Michelson and Morley's notion of a "fixed ether" or a point on the sky from which the Big Bang supposedly began. But, according to *inseparability*, that will never happen. Every portion of the Infinite Universe moves with respect to all other portions. To make our measurements, we are forced to choose a reference point, but it would be foolish to think that reference point to be fixed with respect to other microcosms. Today we assume that no microcosm can exist by itself, and slowly we are beginning to realize that no microcosm can even hold together as a body by itself. In spite of this, it was possible for Newton to ignore such important details and still gain tremendously useful information by emphasizing only one feature of the body as a whole: its motion as a whole.

Remember that *materialism* implies that all phenomena are the interactions of matter with matter. Newton reduced these interactions to two principal types: acceleration and deceleration. That still holds, but if we view the interacting bits of matter as microcosms instead of atoms, then four additional interactions are required. Taken together, the six possible interactions between microcosm and macrocosm constitute the foundation of *neomechanics*. The real interactions from which these abstractions have been derived usually involve all six to varying degrees, but at any moment, one of them usually dominates. The six interactions are:

Acceleration

Deceleration

Absorption of Motion

Emission of Motion

Absorption of Matter

Emission of Matter

Newtonian mechanics is a veritable celebration of the first two interactions: acceleration and deceleration of the microcosm as a whole. Because Newton's microcosms were solid, hard, finite bodies, he could not conceive of the absorption and emission of motion and matter from within those bodies. The contact surfaces on Newton's theoretical microcosms were inflexible and impenetrable.

Today we realize that the interface between microcosm and macrocosm is more like an elastic sieve than a solid wall. It selectively admits motion and matter. This sieve-like interface consists of submicrocosms, no two of which are identical. The upshot is that no two portions of the interface have identical resistances to other matter in motion. The interface itself is neither perfectly inelastic as Newton supposed, nor is it perfectly elastic. The properties of selectivity and elasticity are not unique. Any part of space, any part of the universe, provides some access and some resistance to other matter and the motion of matter. A microcosm is like a house. It may have walls, but it has doors too.

An interface between microcosm and macrocosm may be coincident with a visible structure, as is an eggshell, or it may be invisible, as is the interface between the solar system and the rest of the galaxy. For theoretical purposes, we find that an imagined interface is often suitable and sufficient. It incidentally has the advantage of preventing the contents of the enclosed microcosm from being thought of as totally isolated from the rest of the universe. As a reduction based on the theory of the <u>univironment</u>, one of the primary intents of neomechanics is to make it virtually impossible to conceive of a microcosm without a macrocosm. It is thus an outright rejection of the systems philosophy of the 20th century. It also endeavors to make it virtually impossible to conceive of a microcosm that does not contain other microcosms. It is thus also a rejection of classical mechanics.

As will become more evident as we go along, neomechanics is the foundation of a philosophy that strives to be neither microcosmic nor macrocosmic. A valid criticism of this revision of mechanics is the following: you have perhaps removed Newtonian idealism from the microcosm-macrocosm interaction, but you have merely moved it to the level of the submicrocosm. This is true. In the reduction to follow, submicrocosms and supermicrocosms—the individual parts of the microcosm and the macrocosm—are, at some point, unavoidably treated as rigid, Newtonian bodies. It is true that at some point, we must ignore the non-Newtonian interactions (absorption and emission of motion and matter) between submicrocosms, but, as it turns out, this is of no significant consequence to the explanation.

We can always include the absorption and emission of motion and matter as factors when the focus is on *them*, as microcosms. According to ***infinity***, we are left with no choice; we would have to ignore the internal and external motions of at least some parts of the universe to devise an explanation which, to be expressed, must be finite. As pointed out before, because of ***infinity*** scientists are forced to disregard portions of the universe. With neomechanics, we analyze what we believe to be the "main features" of the microcosm and the "main features" of the macrocosm in relation to each other. We purposely ignore parts of the microcosm and parts of the macrocosm, not forcing ourselves to ignore either *all* of the microcosm or *all* of the macrocosm.

Similarly, one could question the nature of *contact*, in the neomechanical view. With Newton, the rigidity of the interacting bodies left little doubt in the minds of the idealistically inclined as to the nature of the contact. The contact point was clearly the place where the two bodies met. The motion was transferred from whole body to whole body and submicrocosms had nothing to do with the transfer. In neomechanics, on the other hand, the motion is transferred from submicrocosm to submicrocosm and from subsubmicrocosm to subsubmicrocosm, and so on. This view provokes some questions: When does this transfer end and why? What *is* contact? Unfortunately, like other questions involving infinite regressions, this cannot have an answer that would satisfy an absolutist.

One could say that there is not enough "time" for the infinite regression to proceed through an infinite number of microcosms within microcosms. But this again would be begging the question—a primary and necessary characteristic of Infinite Universe Theory itself. With **causality,** we assume an infinite quality to the cause-effect relationship. It is appropriate that neomechanics include this quality at the same point where cause becomes effect in the Newtonian view. The following are short descriptions of the six possible interactions that occur between the microcosm and the macrocosm.

Let me repeat some especially important points. In the descriptions below the primary focus is on the univironment: the microcosm and the macrocosm considered equally. As mentioned, however, we consider both the microcosm and the macrocosm as infinitely subdividable. Again, the divided portions of the microcosm I call _submicrocosms_, and the divided portions of the macrocosm I call _supermicrocosms_. The prefixes to these terms are not intended to imply any special characteristics other than of position relative to the boundary between the microcosm and the macrocosm. In addition, submicrocosms, in particular, seldom have the degrees of freedom noted in the illustrations.

1. Acceleration

When a microcosm is hit by a supermicrocosm, it gains motion (Figure 13). In the ideal model, the object (microcosm) as a whole is accelerated. As mentioned previously, in classical mechanics, the walls of the microcosm are considered perfectly inelastic: an acceleration of the wall at the point of impact produces an "instantaneous," acceleration of the opposite wall as well as the "solid" matter in between. To the degree that the impact is not in a perfectly straight line, some of the acceleration causes the microcosm to rotate—an important fact to be discussed later.

A nice example of acceleration occurs in the physics demonstration commonly known as "Huygens Pendulum" (Figure 14). Releasing the first ball allows it to hit the second ball, transferring its motion as it loses its own motion. This basic interaction occurs in every description of causality. In classical mechanics, which assumes *finity*, we might say that the motion of the first ball is the cause of the motion of the second ball. The equation for the description is F = ma, where m is the mass and a is the acceleration, that is, the change in velocity transmitted from the first ball to the second. Note that force is only a calculation. As mentioned, there are no such things or motions that could be considered "forces"—only microcosms in motion exist. The video shows only microcosms in motion—there are no "forces" to be seen anywhere. Nonetheless, the force calculation is extremely valuable. For instance, we can use it to determine mass by applying a known amount of motion to a microcosm. That gives mass its definition: resistance to acceleration. We use the acceleration produced by gravitation to determine mass.

In neomechanics, however, elasticity is required in the walls of the microcosm for there to be any acceleration. In this view, acceleration results from the displacement of submicrocosms at the point of contact. The motion of the colliding supermicrocosm is transferred across the wall from supermicrocosm to submicrocosm and from submicrocosm to submicrocosm, accelerating each along what would have been the approximate path of the supermicrocosm had there not been a microcosm there.

116	*Infinite Universe Theory*

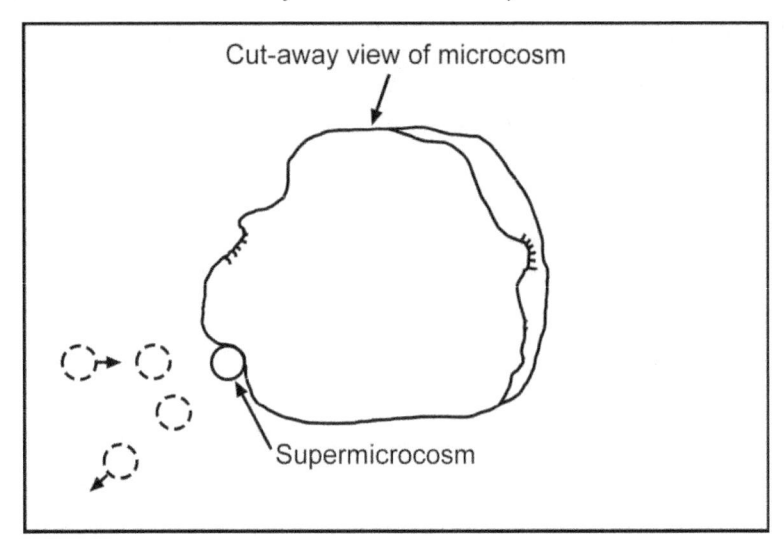

[Figure: Cut-away view of microcosm with Supermicrocosm]

Figure 13 ACCELERATION OF THE MICROCOSM. A relatively high velocity supermicrocosm (external microcosm) collides with and transfers motion to a relatively low velocity microcosm. As a result, the microcosm moves to the right with increased velocity. Dashed lines indicate former locations of the impacting supermicrocosm ("super" because it is outside the microcosm, not because it is anything special).

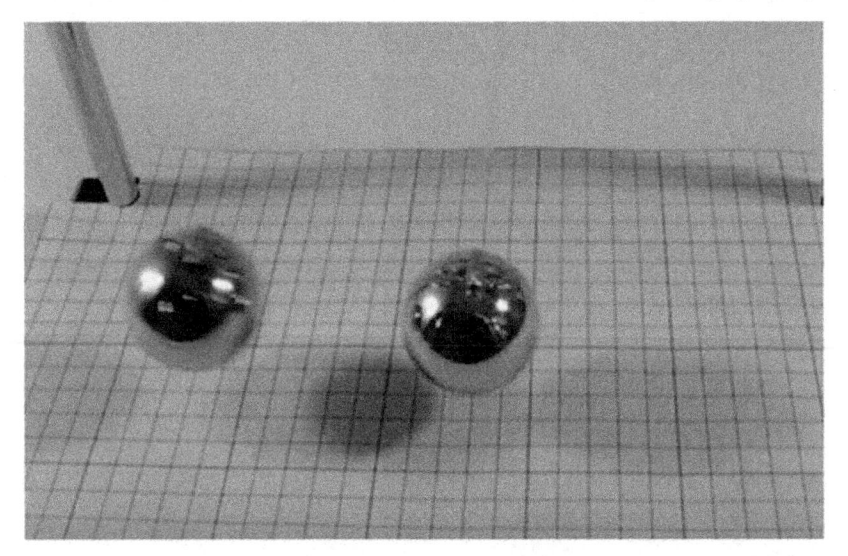

[Figure: Two metal balls on grid paper]

Figure 14 Video showing acceleration and deceleration demonstrated by pendulums at http://go.glennborchardt.com/pendulums. Credit: httprover.

To elaborate on the neomechanical view of acceleration, let us consider the acceleration of a railroad train. In this example, we view the entire train as a microcosm and the individual cars as submicrocosms. When the end car is pushed from behind by another engine (considered here as a supermicrocosm), it subsequently collides with the car in front of it. This in turn, collides with the next, and so on, until the entire train is moving. Motion within the macrocosm is transferred to submicrocosms, which becomes motion of the microcosm as a whole.

2. Deceleration

When a microcosm hits a supermicrocosm, the microcosm loses motion (Figure 15). In billiards, the collider is decelerated so much that it normally stops entirely, while the collidee gains its motion (Figure 14). The microcosm as a whole is decelerated per Newton's Third Law of Motion. Again, this change in motion can be viewed in the classical way in which the trailing edge of the perfectly inelastic body is decelerated at exactly the same time as the leading edge. Alternatively, it can be viewed in the neomechanical way, in which the deceleration is transferred from submicrocosm to submicrocosm, with a slight delay in deceleration of the trailing edge. The microcosm is compressed in the direction of travel and is decelerated as a whole. To the degree that the impact is not in a perfectly straight line, some of the deceleration causes the microcosm to rotate.

A train is decelerated in the opposite way it is accelerated. The change in motion is transferred from car-to-car, with an inevitable, though temporary shortening of the train being the result.

It is unlikely that a person steeped in classical mechanics would choose a train as the best elementary example of acceleration and deceleration. A Newtonian purist would probably choose a single car instead, trying to convince the

118 *Infinite Universe Theory*

reader that the car moves as a whole, with the front moving at *exactly* the same time as the rear. In the neomechanical view, it does not.

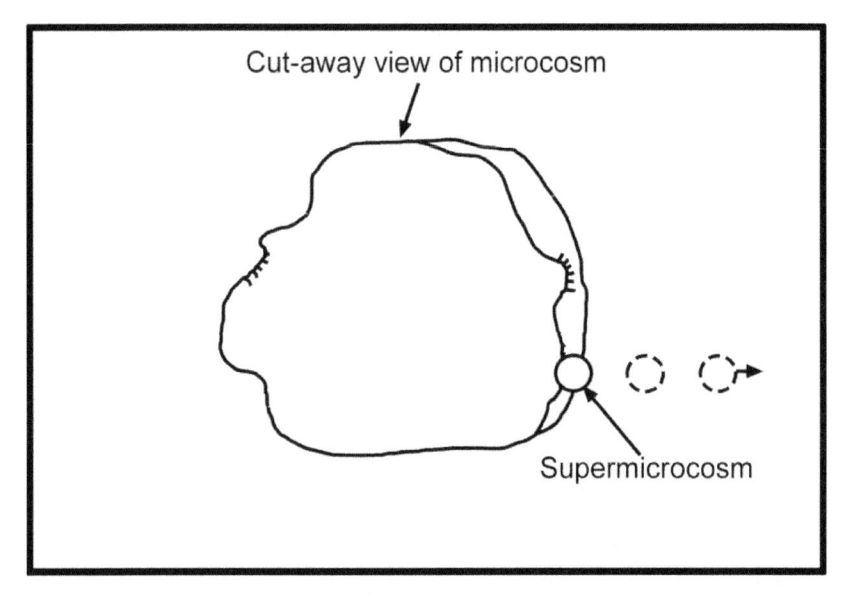

Figure 15 DECELERATION OF THE MICROCOSM. The microcosm collides with and transfers motion to a low velocity supermicrocosm. As a result, the microcosm tends to move to the left losing part of its rightward velocity while the supermicrocosm moves to the right.

Again, neomechanics requires submicrocosms to transfer motion from one portion of the microcosm to another. Classical mechanics ignores submicrocosms or assumes they are in perfect contact, which amounts to the same thing. This is why the Newtonian model must be one of an "ultimate" particle filled with an indivisible substance through which motion can be transferred perfectly and instantaneously.[101] Such imagined particles must be perfectly spherical because, if they were not,

---

[101] Note that much of the evolution of mechanics has included innumerable steps away from the simple idealization presented here for pedagogical reasons. Nonetheless, that evolution has not eliminated the tendency for some regressive physicists to hypothesize the existence of solid matter or finite particles.

any imperfection would betray the presence of the submicrocosms within. As an example, upon cursory examination, the Moon might appear to be a perfect sphere. However, a closer look will reveal its imperfections, with hills and valleys and variations in rock type just like Earth. With our assumption of *relativism,* we predict that every microcosm will display such imperfections—there can be no solid microcosm not composed of individual parts. In other words, the Infinite Universe requires imperfection.

Our adoption of the Theory of the Univironment goes "beyond Newton" because it forces us to admit that the acceleration and deceleration of the microcosm always requires internal interactions. We must conclude that the transfer of motion from microcosm-to-microcosm does not occur in the perfect way assumed by Newton. As a result, there are losses; the transfer of motion cannot be 100% efficient. Some of the motion of the whole appears as motion of the parts, leading to interactions that he did not consider: the absorption and emission of motion, the essence of the *closed system* of today's parlance.

3. Absorption of Motion

Again, microcosms have submicrocosms through which the motion of impacting supermicrocosms is transferred. The motion of acceleration or deceleration is, in effect, temporarily absorbed. It does not instantaneously appear as a change in motion of the microcosm as a whole because the only way that could happen would be for the microcosm to be filled with perfectly solid matter. Such a microcosm would not contain submicrocosms. If ever found in nature, it would be a falsification of neomechanics. It has not been found because perfectly solid matter is only an idea—the ideal end member of the matter-space continuum. Real things always have both properties. That is one reason Finite Particle Theory always will be fruitless—a single particle making up all matter will never be found. Sometimes, the impact against the univironmental interface barely changes the motion of the microcosm as a whole. In such cases, most of the motion of impact is absorbed by submicrocosms as internal motion.

When supermicrocosms hit the microcosm, the elasticity of the interface allows motion to be transferred to the submicrocosms within (Figure 16). The submicrocosms are sped up, and we say that the microcosm has gained internal energy (or enthalpy, H, in the lexicon of thermodynamics). In effect, this submicrocosmic increase in velocity produces an increase in what we would calculate as an increase in submicrocosmic momentum. Submicrocosms with increased momenta impact the inside of the microcosmic boundary at increased velocities. This increased internal motion is measured as an increase in mass of the microcosm as a whole.[102] For example, a hot teakettle has more internal motion and weighs more than a cold one.[103] This is because mass is defined as resistance to acceleration. Impacts due to an increase in internal momentum counteract the impacts needed to accelerate the microcosm when we determine its mass. Look at it this way: Suppose we remove the microcosmic boundary. Then, the submicrocosms would collide with the supermicrocosms we normally use to determine resistance to acceleration (mass). These collisions would decrease the velocity of the supermicrocosms, a decrease that would be attributed to resistance to acceleration (mass). Note that the amount of matter remains the same before and after the absorption of motion. Normally, we think of mass and matter as being identical. This is one important case in which they are not. Remember that "mass"

---

[102] Lewis and Randall, 1923, Thermodynamics.
[103] Gardner, 1962, Relativity for the million. [I like this quote even though Gardner did not explain it well. The upshot is that mass is a function of internal temperature per Maxwell's $E=mc^2$ equation. Einstein's relativity is not required.]

is the resistance to acceleration, while "matter" is an abstraction—a general name we give to things. In neomechanics, we assume that microcosms must have mass because every microcosm must contain submicrocosms in motion. Contrary to Einstein, we would never propose the existence of a massless particle.

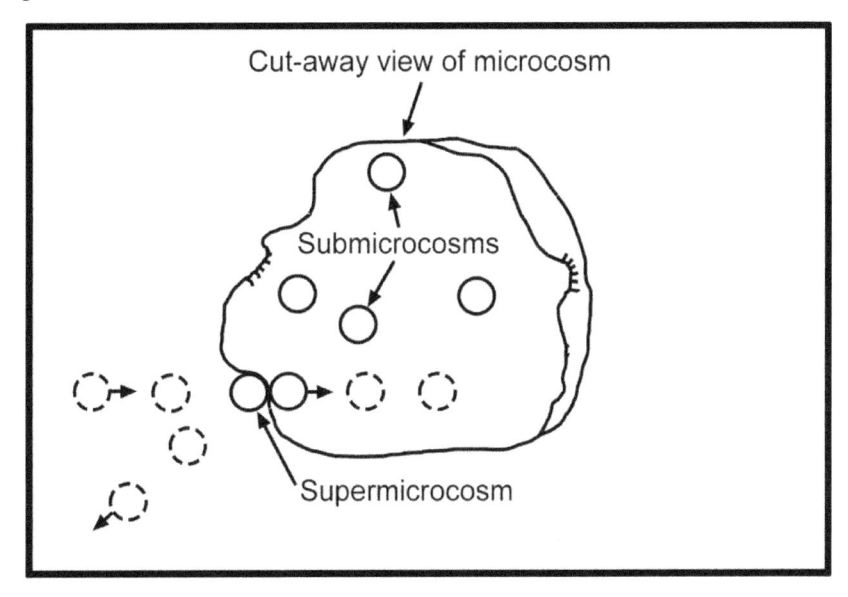

Figure 16 ABSORPTION OF MOTION. A high-velocity supermicrocosm collides with and transfers motion to a low-velocity submicrocosm (internal microcosm). As a result, both the microcosm and the submicrocosms inside it are accelerated slightly to the right.

The above examples are idealizations. In reality, they describe only tendencies. Not all of the motion of a converging supermicrocosm can be absorbed internally. For one thing, the submicrocosms are not as free roaming as depicted for the ideal model (Figure 16). As mentioned under acceleration, some of the motion of the supermicrocosm is transferred from submicrocosm-to-submicrocosm in the general direction in which the supermicrocosm was traveling. Submicrocosms to one side of this path tend to be accelerated less than those directly on the path. In a way, the submicrocosms along the path (dashed submicrocosms) tend to move along a "right line," sometimes as a unit, or as a whole. This is precisely the type of motion we idealized for acceleration in which the microcosm as a whole was accelerated.

Impacts from supermicrocosms produce both acceleration and absorption of motion. Acceleration and absorption models are idealized, nonexistent end members of a continuum, which Newton reduced to one end member: acceleration. Because his model lacked submicrocosms, he was able to neglect the absorption part of the continuum in presenting his three laws of motion.

There are a few additional important details concerning the absorption of motion. In the treatment above, we assumed a microcosm of constant volume, but no real microcosm could have a perfectly inelastic interface for maintaining a constant volume. Again, the increased internal motions of the submicrocosms are likely to impact the interior surface of the microcosmic boundary with increased momenta, pushing it outward toward the macrocosm. In short, there will be a tendency for the volume to increase—for the microcosm to expand. The motions of the submicrocosms are then spread over a larger volume, the density decreases, and entropy increases (a divergence or decrease in order) as the submicrocosms diverge from each other. In addition, it is possible for internally absorbed motion to appear as rotational motion of some of the submicrocosms.

Hammer and Nail

The absorption of motion by the microcosm is illustrated by the interaction between the hammer and the nail (Figure 17). The microcosm of the nail is accelerated by the supermicrocosm of the hammer, but if the wood is especially hard, the nail absorbs much of this motion instead. The submicrocosms within the microcosm of the nail and the supermicrocosm of the hammer are accelerated and both the hammer and the nail become hot. Instead of appearing as motion of the whole, much of the transferred motion appears as motion of the parts.

Figure 17 The collision between a hammer and a nail cause both to become warm. Credit: Mark Wasteney at flickr.com.

4. Emission of Motion

The emission of motion (Figure 18) is the opposite of the absorption of motion. Emission occurs when a rapidly moving submicrocosm collides with a slower supermicrocosm. This transfer of motion results primarily in deceleration of the submicrocosm and acceleration of the supermicrocosm.

While the absorption of motion increases the momenta of the submicrocosms within the microcosm, the emission of motion decreases them. Mass decreases and entropy increases because of the emission. And because there will be fewer impacts on the interior of the univironmental interface, the volume will tend to decrease. This secondary reaction increases density and decreases entropy (a convergence or increase in order), offsetting some of the effects produced by the initial reaction. Of course, it is also possible for rotational motion of the whole to be lost through emission.

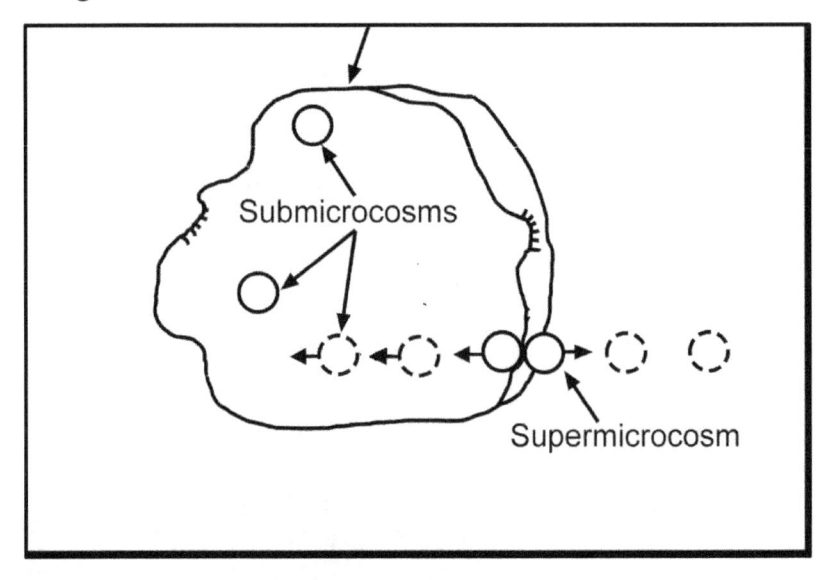

Figure 18 EMISSION OF MOTION. A high-velocity submicrocosm collides with and transfers motion to a low-velocity supermicrocosm. As a result, both the microcosm and the submicrocosms inside it are decelerated slightly to the left.

Water and Frying Pan

Emission is complicated, but a simple example illustrates its principal results. Consider what happens when cold water is dropped on a hot frying pan. The microcosm of the hot frying pan has within it submicrocosms, atoms, that have rapid vibratory motion. Upon contact, some of this motion is transferred to the more slowly vibrating submicrocosms of water molecules within the supermicrocosm of the water droplet. The water molecules absorb the motion emitted by the hot frying pan. The internal motion of the submicrocosms within the droplet increases—the temperature rises. Some of this submicrocosmic motion appears as an acceleration of the supermicrocosm as a whole. We observe this as the tendency for the droplet to assume a spherical shape and, if the frying pan is hot enough, to leave the surface, propelled by the extremely rapid motions of the vapor that forms beneath it.

Atomic Fission and $E=mc^2$

Here is where you become smarter than Einstein. The world's most famous equation is used to calculate the effect of the emission of motion. The equation is simply the bidirectional equivalent of the one we use to calculate kinetic energy: $KE=1/2\ mv^2$. During fission, atomic particles fly in all directions, with those directions averaging out to be both forward and backward, yielding twice the energy that we calculate from a simple forward motion. We use "c" instead of "v" because the emission occurs at the speed of light. As in the cooling of the teakettle, the momenta of the submicrocosms within the microcosm decrease as they collide with supermicrocosms across the microcosmic boundary. As a result, the mass of the microcosm decreases as the emitted motion is transmitted to the macrocosm (Figure 18).

Now, as an aether denier, the young Einstein had an obvious problem at this point. His macrocosm of concern theoretically lacked supermicrocosms to which motion could be transferred. It was perfectly empty space. The analysis begged for a medium containing the supermicrocosms responsible for light transmission—that is why neomechanics requires aether. Although completely in accord with **conservation**, the idea that this job might be performed by unseen aether particles was anathema to the supremely idealistic Einstein. So he devised an even more fantastic solution: a massless wave-particle later dubbed a "photon," which became the personification of energy. In this regard, the $E=mc^2$ equation has been widely misinterpreted as representing the "conversion of mass into energy" and the belief that mass and energy are the same thing. That is not possible because energy neither exists nor occurs. As

I mentioned previously, energy is a helpful calculation, a matter-motion term multiplying a term for matter times a term for motion. Nevertheless, it is not matter and it is not motion.

Most importantly, the submicrocosmic motion within the fissioning atom cannot appear as an object fleeing the scene of the crime as some kind of phantasmagoric particle of energy traveling through perfectly empty space. Again, the $E=mc^2$ equation is merely a calculation involving the transformation of one kind of motion into another kind of motion, just like any other.

Because motion and matter are inseparable aspects of a single reality, the next two interactions are special cases of absorption and emission, only this time the interactions involve only matter, the essence of the *open system* of today's parlance. The effects of the absorption of matter or motion are similar. Both, for example, are convergences, and thus both result in increases in mass and density, and decreases in entropy (a convergence or increase in order). Note that matter cannot be absorbed or emitted without accelerating or decelerating the microcosm as a whole.

5. Absorption of Matter

The sieve-like character of the interface between the microcosm and the macrocosm allows a converging supermicrocosm of an appropriate size and velocity to penetrate the interface and enter the larger microcosm (Figure 19). This addition of a smaller microcosm to a larger one obviously results in a net increase in mass and, to the extent that the volume of the microcosm does not increase, results in an increase in density. At the same time, this amounts to a convergence. When microcosms come together like that, they have the opportunity to produce an increase in order. In thermodynamic parlance, this is known as a decrease in entropy. Entropy increases when things come apart per the Second Law of Thermodynamics. Obviously, when things come apart, the order produced by the prior juxtaposition will tend to be destroyed.

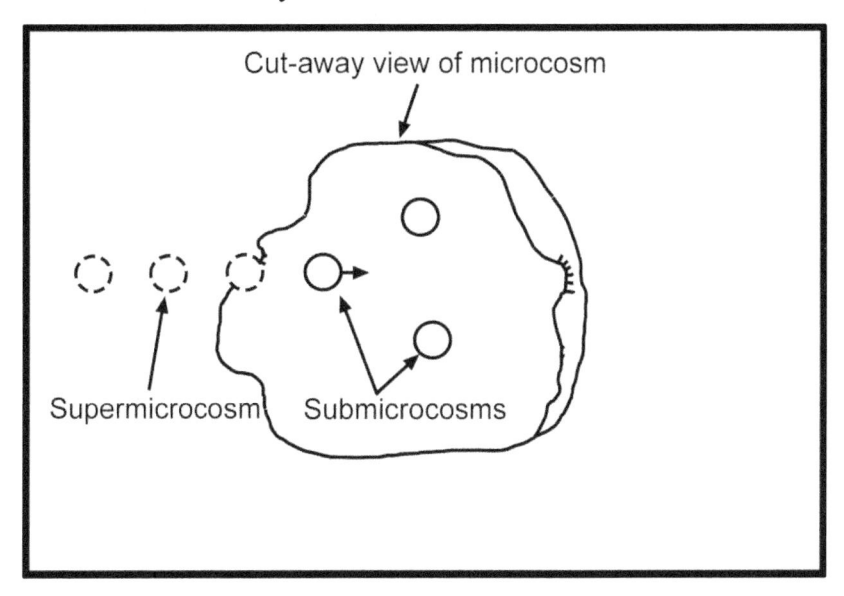

Figure 19 ABSORPTION OF MATTER. A supermicrocosm enters a low velocity microcosm and becomes a submicrocosm. The supermicrocosm also could be absorbed on the surface of the microcosm, thus widening the microcosmic boundary.

Filling a Container

Microcosms have openings that allow matter to enter from the macrocosm. The filling of a container with water is a good illustration of the absorption of matter. The small size of the supermicrocosm of the water allows it to enter the microcosm of the container when there is a convergence between water and container. In addition, the microcosm of the water accelerates the microcosm of the container slightly, while the water becomes decelerated, remaining inside the container.

6. Emission of Matter

The same characteristics of the interface that allow the absorption of supermicrocosms may also allow the emission of submicrocosms (Figure 20). The submicrocosms within a microcosm have their own inertial motions that result in their continual bombardment of the interface. Thus under certain conditions, a submicrocosm may break through the interface and leave the microcosm under its own inertial motion. In this regard, it is indeed unfortunate that the available terms for describing this process, such as *emit, eject,* and *release,* have teleological connotations that fail to emphasize the inertial aspects of this motion. Remember that teleology occurs when we ascribe will or purpose to a natural process. However, in neomechanics motion is always inertial, even those involved in the collisions we recognize as causes per the Second Law of Motion. Submicrocosms that appear to have been emitted or ejected do so under their own inertia. The words (emitting and ejecting) we use to describe that process are unfortunate. Actually, there never is anybody or anything that does the emitting. Instead, it is analogous to what happens when the gate to the lion's cage is opened. The lion was not emitted; he just plain left.

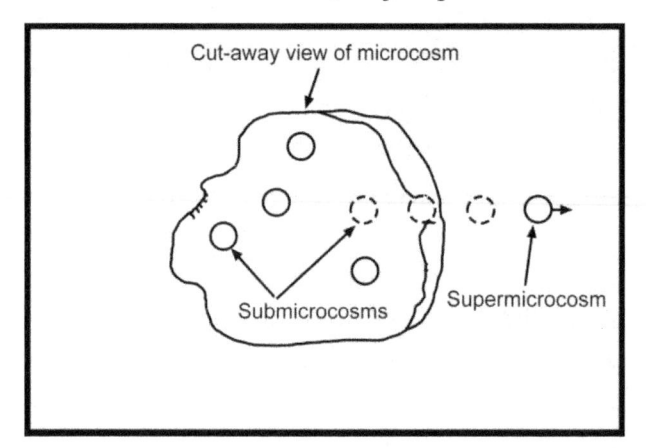

Figure 20 EMISSION OF MATTER. A submicrocosm leaves a low velocity microcosm and becomes a supermicrocosm.

The effects of the emission of matter are just the opposite of its absorption. The mass decreases, and to the extent that the volume of the microcosm does not decrease, there is a decrease in density and an increase in entropy (a divergence or decrease in order) as well. There will be some decrease in volume, because with fewer submicrocosms to impact the interior of the interface, the macrocosm, with its continual bombardment of the interface, will encroach on the microcosm.

Boiling Water

Boiling water is an example of the emission of matter. For this to occur, the motion of the water molecules (submicrocosms) must be increased by the application of heat to the microcosm (kettle). Eventually, some of the water molecules will be ejected as water vapor, becoming supermicrocosms with respect to the microcosm.

This divergence from the microcosm amounts to an increase in entropy (a divergence or decrease in order). It is this type of interaction we celebrate in the classical demonstration of the Second Law of Thermodynamics (Figure 21).

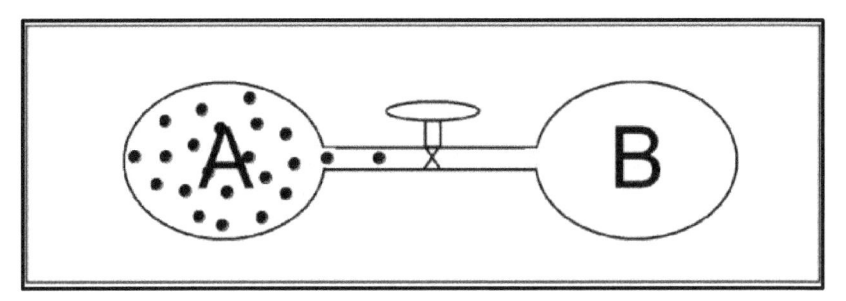

Figure 21 Demonstration of the Second Law of Thermodynamics, Newton's First Law of Motion, and the tendency for microcosms to move toward univironmental equilibrium. Opening the valve will allow the gas molecules in A to move into vacuum chamber B under their own inertia.

As mentioned, the six ideal interactions described here are never found in pure form. Any actual interaction between microcosm and macrocosm must involve all six interactions to varying degrees. Thus, a converging supermicrocosm: 1) accelerates the microcosm as a whole, 2) produces absorption of motion internally, and 3) penetrates the microcosmic boundary, at least temporarily adding some of its matter to the microcosm. Moreover, as Newton observed, but stated in a different form, the microcosm and macrocosm undergo equal and opposite reactions per conservation. Thus, a converging supermicrocosm: 4) decelerates, 5) emits some of its internal motion, and 6) loses matter to the microcosm. Newton's Third Law of Motion only includes deceleration.

Univironmental interactions are inherently dialectical. There can be no acceleration of the microcosm without a corresponding deceleration of the macrocosm and there can be no absorption of motion or matter within the microcosm without a corresponding emission of motion or matter from the macrocosm. The microcosm and the macrocosm undergo equal and opposite irreversible reactions. Not only are their momenta and positions changed as a whole, but the momenta and positions of their contained submicrocosms and supermicrocosms are changed as well. An interaction between the microcosm and the macrocosm irreversibly changes both.

We must always remember that these neomechanical idealizations of univironmental interactions are, for the most part, gross oversimplifications. In reality, each microcosm undergoes an infinite number of these interactions with the macrocosm at every moment. The interactions, in fact, are what define the microcosm. Without the differences in the motions of matter on either side of the univironmental boundary, we could not even discern a microcosm. We would have no reality on which to base our imagined model, which by comparison, is a mere cartoon of the infinitely complicated microcosms we encounter every day. The neomechanical reduction nevertheless assimilates some powerful assumptions requiring a considerable revamping of conventional scientific explanations. For instance, this scheme requires no "attractive force." Remember that force is F=ma, a calculation describing the collision of one microcosm with another. There can be no "attractive force," as Newton admitted when he proposed that gravitation was a push, not a pull.[104] In addition, univironmental determinism and neomechanics do not treat motion or time as a fourth dimension of material objects. The expansion of one portion of the Infinite Universe invariably occurs at the expense of other portions of the universe. Because neomechanics assumes *infinity*, and the impossibility of nonexistence, the concept of the universe expanding into empty space makes no sense.

Striking at the heart of systems philosophy, neomechanics reminds us that microcosms are neither perfectly isolated nor perfectly nonisolated. When we imagine them to be nonisolated, the decrease in entropy resulting from the inevitable convergence coming from the macrocosm is just as "spontaneous" as the increase in entropy undergone by supposed isolated systems as their parts diverge into the macrocosm. As we have seen, microcosms are always increasing or decreasing in mass, velocity, density, volume, entropy, and apparent order. I write "apparent" because order is a subjective concept. The bias is this: When things come together, we generally consider them more orderly; when things come apart, we generally consider them more disorderly. Objectively, neither condition is more favorable than the other is.

At times, the motion of matter within the microcosm is such that the macrocosm yields on nearly all fronts—the microcosm

---

[104] Newton, 1718, Optics.

expands. At other times the motion of matter within the microcosm is less and the macrocosm pushes in from nearly all sides—the microcosm is compressed. The macrocosm yields to a certain extent and resists to a certain extent. This univironmental relationship determines the spatial extent of the microcosm. Since all things either converge on or diverge from other things at all times, the interface between the microcosm and the macrocosm moves back and forth. Like a beating heart, the microcosm pulsates with the macrocosm, expanding and contracting—demonstrating why matter must be in constant motion. The space-time position of the univironmental boundary is determined neither by the microcosm nor by the macrocosm, but by both in a reciprocal relationship. Even the collision of one microcosm with another always takes time (involves motion). The conversion of convergent motion into divergent motion always takes time—however short. Nothing takes place in zero time.[105] In particular, no "equilibrium" is evidence for the absence of motion.

Each microcosm moves through the macrocosm in whatever direction that yields to it. This it does until, inevitably, it reaches a part of the macrocosm where the resistance, the motion of matter, is comparable to that of the microcosm. The microcosm is slowed, its motions becoming more like those of the macrocosm. Shortly, the microcosm and the macrocosm display an approximate unity, a sort of ephemeral truce along the univironmental border. Such a microcosm has moved toward *univironmental equilibrium*, the subject of the next chapter.

---

[105] Borchardt, 1978, Catastrophe theory. [This was my first formal incursion into scientific philosophy. It was a letter to the editor debunking the implication in catastrophe theory that changes could occur instantaneously or in zero time.]

Neomechanical Conclusion

The inclusion of ***infinity*** in our assumptive foundation has resulted in a new way of looking at the universe. A few simple comparisons between classical mechanics, neomechanics, and relativity reveal stark differences beyond Infinite Universe Theory's claim that the universe had no beginning and will have no end (Table 5). With ***infinity***, we were able to dismiss idealistic claims about the reality of solid matter, empty space, and nonexistence. The neomechanical assumption that matter is infinitely subdividable and continually in motion implies that the number of causes for an effect is infinite and that there are no true constants in nature.[106] Because we assume time to be motion, it cannot be a dimension and cannot dilate. Neomechanics uses Newton's three laws of motion, which describe only pushes in nature. We found it unnecessary to hypothesize an "attractive force." On the other hand, we agree with classical mechanics that aether is theoretically necessary for non-contradictory explanations in physics. Unlike relativity, neomechanics considers space-time to be a matter-motion term that is a conceptualization, not a reality. The object of concern in classical mechanics was the "body;" for relativity, it was the "system." For neomechanics, it is the "microcosm," which demands its counterpart, the "macrocosm" in emphasizing interactions between the two portions of the universe.

Some other differences were discussed in "The Scientific Worldview," namely that the universal mechanism of evolution is univironmental determinism (What happens to a portion of the universe is determined by the infinite matter in motion within and without), not neo-Darwinism, which is only limited to biological systems. The general philosophy of classical mechanics was mechanism and that of relativity is systems

---

[106] We discussed this in detail in "Universal Cycle Theory."

philosophy. The general philosophy of neomechanics is the same as its universal mechanism of evolution: univironmental determinism.

Table 5 Characteristics of classical mechanics, neomechanics, and relativity.

Characteristic	Classical Mechanics	Neomechanics	Relativity
Fundamental assumption	Finity	Infinity	Finity
Is solid matter possible?	Yes	No	Yes
Is a fundamental particle possible?	Yes	No	Yes
Is empty space possible?	Yes	No	Yes
Is nonexistence possible?	Yes	No	Yes
Number of causes for an effect	Finite	Infinite	Finite
Is time a dimension?	No	No	Yes
Number of dimensions in nature	3	3	4+
Can time dilate?	No	No	Yes
Are there constants in nature?	Yes	No	Yes
Is there an attractive force in nature?	Yes	No	Yes
Is there an aether?	Yes	Yes	No
Does space-time exist?	No	No	Yes
Is the universe infinite?	Maybe	Yes	No
Did the universe have a beginning?	Yes	No	Yes
Mechanism of evolution	Neo-Darwinism	Univironmental Determinism	Neo-Darwinism
Philosophy	Mechanism	Univironmental Determinism	Systems Philosophy
Object of concern	Body	Microcosm	System
Action of concern	Motion	Motion	Energy

Chapter 12
Univironmental Analysis

Neomechanics described what happens when matter interacts with matter. That was fine, but, like Newton's laws of motion, neomechanics still might be criticized for not being able to explain *why* things happen. For instance, why was a particular collider headed toward a particular collidee and not some other potential collidee? Scientific theories are supposed to make predictions; observations alone are not enough. In this chapter, we come close to answering the why question, which is an essential part of every prediction.

When not receiving significant collisions from supermicrocosms, all microcosms move toward *univironmental equilibrium*—the slowest motion possible. This is because microcosms cannot speed up by themselves. In Newton's First Law of Motion, the body traveling through empty space on its own inertia does not suddenly get an engine that could speed it up. Thus, a bullet falls to the ground as it encounters gravitational and atmospheric resistance. In other words, if a microcosm collides with anything, it will slow down. That is inevitable in the Infinite Universe, because there is always something there to slow it down. In addition, because the universe is infinite, there always will be supermicrocosms that will tend to speed it up. Thus, in neomechanics, equilibrium clearly refers to the dynamics of the univironment, the relationship between microcosm and macrocosm—the eternal slowing down and the eternal speeding up—the "why" for all events. A *stable* univironment is one in which the motions of the microcosm and the motions of the macrocosm are similar—both must have constituents with similar velocities. An *unstable* univironment is one in which the motions of the microcosm and the motions of the macrocosm are dissimilar—both must have constituents with different velocities. Stability and instability are ideal end members of a continuum. Like the related ideal concepts of absolute rest and absolute motion, neither can exist in reality. The upshot is that equilibria are always temporary. In the Infinite Universe, every microcosm is always in motion. Again, there are no constants in nature. Per *uncertainty*, no two measurements ever give the same result at the limits of measurement, which reflects the infinite subdividability of matter.

As explained under *inseparability* and *complementarity*, one of the failings of systems philosophy is its frequent ambiguity regarding the referent in discussions of stability and equilibrium. Systems often are said to be "stable" or "unstable," as though stability could be a property of the system itself. The error often arises through a conceptual confusion of system boundaries and subsystem boundaries. Thus, two parts of a system may be in relative equilibrium with respect to each other. One part, considered as a microcosm, is stable relative to the other part considered as a macrocosm, and vice versa. This correct observation concerning the two parts of the system is then mistakenly attributed to the system taken as a whole. The true referent, the macrocosm, disappears in a way worthy of the subtlest solipsism. Nevertheless, no matter how stable two parts of a microcosm may be in relation to each other, their combination as a microcosm cannot be permanently stable. That is because the macrocosm always contains supermicrocosms that are present or not present, with either situation likely to disrupt that stability.

In spite of its many anthropocentric concepts, such as force and attraction, classical mechanics availed scientists of a generally objective image of the world. The inertia concept, for example, did not require an answer to why things moved in the first place. Once in motion relative to other things, Newtonian objects continued to move—*unless* they collided with other things. The results of these collisions, in turn, were then considered independent of a willing agent. Moreover, as long as scientists did not apply classical mechanics to living things, the question of goal or purpose remained in the background.

Nevertheless, the naturalistic picture courted by the Newtonian reduction based on *finity* encouraged attempts to discover an equally naturalistic explanation for what appears to be goal-oriented behavior. The first of these was formulated by French mathematician and philosopher P.L.M. de Maupertuis (1698-1759) as the Principle of Least Action.[107] Based on the Newtonian tradition, the Principle of Least Action emphasized the acceleration and deceleration of whole bodies. Focusing on one of two bodies, it stressed that when a high-velocity body overtakes and collides with another, it always slows down as a result. Similarly, a body never speeds up on its own. Thus, the motion or "action" of a body either remains constant or tends to decrease—as long as it travels through "empty space"[108] or encounters resistance from slower bodies. The principle was reintroduced in the 19th century by Hamilton[109] and even was extrapolated to sociology in the 20th century by G.K. Zipf.[110]

The demise of classical mechanism also led to the decline in popularity of the Principle of Least Action. Today, naturalistic explanations of goal-oriented behavior generally rely instead on the Second Law of Thermodynamics, misinterpreted, of course,

---

[107] Maupertuis, 1751, Essai De Cosmologie; Jourdain, 1913, The Principle of Least Action.
[108] I often use quotes around terms representing ideals. As mentioned before, "empty space" is an idea or illusion. Only naïve idealists could believe perfectly empty space exists. Here, it is used to represent a portion of the universe containing matter that produces insignificant resistance to the motion of other matter.
[109] Famous Irish physicist (1806-1865) who revamped classical mechanics.
[110] Zipf, 1949, Human Behavior and the Principle of Least Effort.

from the one-sided viewpoint of systems philosophy.[111] The Principle of Least Action was born with similar problems.

Actually, the distinction between classical mechanism and systems philosophy was never as great as I generally have portrayed it. The underpinnings of systems philosophy have been with us since the utterance of the first anthropocentrism. In astronomy, for instance, the Ptolemaic system was nothing if not microcosmic. As I pointed out before, the traditional orientation was even built into Newton's First Law. Per the indeterministic assumption of *finity,* a body was said to travel in a straight-line *unless* it encountered another body, not *until*.

Being derived from the First Law, the Principle of Least Action carried with it the seed of systems philosophy and its antidialectical concentration on the *collider* rather than the *collidee*. Rapidly moving bodies inevitably were slowed down when they collided with other bodies and thus the colliders always tended toward *least* action. Nevertheless, other mechanists could not avoid seeing that moving bodies inevitably were speeded up when faster moving bodies hit them. Collidees invariably exhibited *most* action rather than *least* action as their "goal"—a phenomenon wreaking as much confusion in classical mechanism as it does today in systems philosophy.

The Newtonian penchant was for considering space as fixed and for considering the first body as primary and inevitable and the second as secondary and accidental. This provided the solipsistic excuse for choosing least action rather than most action as the goal toward which things supposedly were headed. Of course, relative to the motions of surrounding bodies, the actual motions of real bodies range between rapid and slow. Like the Second Law of Thermodynamics, which succeeded it, the

---

[111] Borchardt, 2008, Resolution of the SLT-order paradox.

Principle of Least Action was useful only for bodies whose motions were rapid relative to their surroundings, that is, they had to be relatively isolated from their surroundings. Without a certain ignorance of the macrocosm, the Principle of Least Action was worthless.

A deterministic explanation of the appearance of goal-oriented behavior could be discovered only from the univironmental point of view. Newton was moving in this direction when he admitted the possible existence of the second body in his First Law of Motion. Of course, depending on which of the two colliding bodies one focused, action after contact could be either at a minimum or at a maximum. Even though Special Relativity Theory implied there was no basis on which to choose between these two possibilities, tradition continued to favor the Newtonian bias.

From the systems point of view, there was a convenient way interactions could be made to fit the Principle of Least Action. Whenever the "action" of a body was observed to increase instead of decrease because of a particular interaction, one could simply expand the boundaries of the system to include the impacting body. When this was done and the rest of the macrocosm was ignored, the amount of action after the impact would still be a minimum and the rule would be preserved. As we have seen, this is the approach used in current interpretations of the Second Law of Thermodynamics—minus the mechanical imagery. In the end, both the Principle of Least Action and the Second Law of Thermodynamics apply only to such ideally isolated systems and say nothing whatever about the macrocosm. The Principle of Least Action was replaced by the Second Law of Thermodynamics partly because Least Action furnished an inadequate view of the microcosm in spite of its microcosmic focus. Following Newtonian tradition, its microcosms tended to be platonic forms—wholes without parts. Accordingly, the principle ignored the absorption and emission of matter and motion. The objects of concern were finite and therefore the causes of the motions of these objects were considered finite too.

With the Theory of the Univironment, we explicitly acknowledge the dialectical interplay between the infinite microcosm and the infinite macrocosm. Both the microcosm and the macrocosm contribute equally to the motions of the microcosm. Make no mistake, that does not mean that *everything* in the microcosm or macrocosm can be involved in any particular acceleration. The most important causes generally are proximal, that is, they are nearby. The effects of distant motions are governed by intervening transmission velocities. Thus, the cause of a home run is the bat hitting the ball. That might be influenced by local conditions involving wind, humidity, and temperature, but it could not be influenced by current conditions on the Sun—which take eight minutes to arrive on Earth. The upshot is that the microcosm moves through the macrocosm under its own inertial motion, but it does so only to the degree that the macrocosm does not resist this motion. In the univironmental view, the relative resistance of the macrocosm is no less important than the relative inertia of the microcosm. Again, this resistance is proximal, with tiny and distant causes being insignificant.

By studying the univironmental relationship, we try to predict the motions of the microcosm. The possibility of prediction arises because we know that rapid motions within the macrocosm produce rapid motions in the microcosm and *vice versa*. Similarly, slow motions in one absorb fast motions in the other, producing slow motions in the other. Thus, these motions tend to become similar on either side of the univironmental boundary. The upshot is that, at all times, the microcosm approaches equilibrium with its macrocosm, that is, it tends to move toward *univironmental equilibrium*.

Univironmental equilibrium, thus, is the "goal" toward which all behavior is directed. Behavior in general is *not* directed toward least action or most action or toward the most entropy or the least entropy. Instead, the direction of movement of the microcosm is determined in each instance by the relationship between the matter in motion within and without. The microcosmically oriented laws devised by classical mechanism as well as systems philosophy apply only to ideally *isolated* microcosms. From this one-sided viewpoint, the "goal" of ideally isolated systems is toward the most entropy, the greatest disorder, or the greatest divergence of their submicrocosms. From the opposite, equally one-sided viewpoint, the "goal" of ideally *nonisolated* systems is toward the least entropy, the least disorder, or the greatest convergence of their submicrocosms. As with all ideals, neither of these two could possibly exist. But as mentioned under **complementarity**, real microcosms exist throughout the range between these two ideals. Isolation reflects the passivity of a particular macrocosm, as would be the case if the microcosm could exist in perfectly empty space. Nonisolation reflects the activity of a particular macrocosm, as would be the case if the microcosm were continuously bombarded by supermicrocosms.

Thus a microcosm surrounded by a passive macrocosm exhibits "least action" and increasing entropy (a divergence or decrease in order) as its apparent "goal," while a microcosm surrounded by an active macrocosm exhibits "most action" and decreasing entropy (a convergence or increase in order) as its apparent "goal." Earth is a good example. During the night, Earth emits motion as it cools down; during the day, Earth absorbs motion as it heats up. Biological microcosms respond accordingly. Thus, if night continued indefinitely, plants would die, becoming increasingly disordered with their various submicrocosms diverging, becoming scattered throughout the macrocosm. If daylight continued indefinitely, plants would flourish, becoming increasingly ordered with their various submicrocosms converging, joining together to form beautiful microcosms of infinite variety. Of course, plants and animals have adapted to the night/day scenario, evolving rest mechanisms that have proven beneficial for survival.

It has long been known that the concept of equilibrium pertains to the fundamental characteristics of the universe. The question arises as to why the word *univironmental*, must be appended to the word *equilibrium*. Remember that systems philosophy, by definition, views equilibrium as something "in the system." For the modern indeterminist, equilibrium is, at most, a balance between two parts of a system. However, as soon as we view equilibrium as a relationship between microcosm and macrocosm, we leave systems philosophy and indeterminism behind. It is true that when equilibrium is defined properly, the appending of *univironmental*, is merely redundant. Until then, however, the combination of *univironmental* and *equilibrium* serves to remind us of the proper focus. When we say that a particular microcosm is moving toward univironmental equilibrium, there should be no doubt that it is headed for the type of motion that results from the most important motions of the infinity of parts within and without. The nature of the interaction of microcosm and macrocosm is dependent on only two things: the nature of the microcosm and the nature of the macrocosm. This line of thinking leads to another important consequence concerning the interpretation of **causality**. With both the microcosm and the macrocosm containing an infinite number of submicrocosms and supermicrocosms and with the degree of passivity as well as the degree of activity contributing to the result, we must assume that *exactly half of the causes of a particular interaction lie within the microcosm and half lie within the macrocosm.*

The univironmental approach is just an extension of Newton's First Law of Motion, which I mentioned, turns out to be the law of the universe. Newton's body moves for only two reasons: 1) it is in motion and 2) there is nothing stopping it. The second reason is just as important as the first. Without his assumption of empty space, nothing could be said about the motion of his assumed body. Without both, there could not have been a First Law of Motion.

Examples of Motion toward Univironmental Equilibrium

Auto Traffic

The general tendency of microcosms to move toward univironmental equilibrium may be seen wherever we look. Take for example the motion of an automobile in heavy traffic. There are many activities occurring in heavy traffic, but here we are only interested in the motion of one automobile, so we choose it as our "microcosm." The microcosm of the auto entering the fast lane moves toward "univironmental equilibrium" by accelerating to the speed of the vehicles in the surrounding "macrocosm." As long as the vehicles ahead and the vehicles behind stay in their lanes and travel at velocities similar to that of the microcosm of the auto, no collisions between vehicles will occur. The motions within the microcosm and those within the macrocosm are such that the microcosm of the auto has achieved temporary univironmental equilibrium. This situation changes as soon as a change in velocity occurs in the microcosm (the auto) or in the macrocosm (the vehicle ahead or the one behind). The microcosm will be speeded up if it is impacted from behind or slowed down if it collides with the vehicle ahead. That either of these events does not happen more often is, of course, dependent on the motions within the univironment in a complex way.

Impending collisions are routinely avoided because of the reciprocal relationship between microcosm and macrocosm. The driver within the microcosm of the auto may take corrective action to achieve univironmental equilibrium, which normally means the avoidance of an accident. Even if the driver of the microcosm of the auto uses bad judgment, not all may be lost, for the drivers within the macrocosm still may react to this "incorrect" motion in an appropriate way. A change in motion within the microcosm results in a change in motion in the macrocosm and vice versa. A fast microcosm within a slow macrocosm soon becomes a slower microcosm. A slow microcosm within a fast macrocosm soon becomes a faster microcosm.

We can use the above analysis to make the rather trivial prediction: Most autos will travel safely to their destinations. Of course, some will not, and without knowing more about the microcosms involved (the state of the equipment and of the drivers), we cannot predict which ones will not make it. To improve our prediction, we need to consider a few more of the infinite number of factors involved: alcohol, bad tires, bad road, depression, etc. None of this makes it possible for us to make perfect predictions, but we know where to look for improvements: the microcosm and the macrocosm.

Gas Molecules

Gas molecules also undergo motions that demonstrate the tendency for microcosms to move toward univironmental equilibrium. Here, we are only interested in the specific behavior of the microcosm of gas molecules. Our question of the day is "What will they do if we change the macrocosm in a certain way?" As seen in the demonstration of the Second Law of Thermodynamics, the microcosm of the gas-filled chamber loses gas molecules to the macrocosm of the empty chamber when the valve between them is opened (Figure 21). In the absence of suitable material constraints within the macrocosm, the inertial motion of the gas molecules carries them to the region of lesser density of matter of that type. In every case, the motions of matter within the microcosm become more in tune with those of the macrocosm, but in so doing, the macrocosm is unavoidably and irreversibly changed as well.

Controlling the Univironment

Again, it must not be thought that the microcosm and the macrocosm could ever achieve some sort of permanent univironmental equilibrium. This clearly would be impossible because the univironment contains an infinite number of things in constant motion. That is why we need to establish "controlled" conditions when we do experiments. Submicrocosms, being themselves infinitely varied, converge on and diverge from each other, continually evolving new types of combinations and new types of dissolutions that eventually disrupt the equilibrium from the inside. Supermicrocosms undergo similarly irreversible reactions, producing new combinations and dissolutions that eventually disrupt the equilibrium from the outside.

As an example, suppose we want to test the effectiveness of a particular soil amendment for growing a particular crop. The easiest test is to prepare two plots on a uniform soil, one with the amendment and one without the amendment. Sure, there are an infinite number of factors in growing crops, but here we have changed only one. We will have our answer if we get a better crop with the amendment than without. If we repeat the experiment, because of ***infinity***, we will surely get a slightly different result. But if the difference between the "with" and "without" plots is large, we will consider the factors accounting for the variations between repeats of the experiment to be insignificant.

The key to all this is to understand the univironmental equilibrium at hand and to change only one factor at a time, be it macrocosmic (fertilizer) or microcosmic (seed variety). Multivariate tests sometimes need to be performed, but these tend to complicate the results proportional to the number of variables involved. As mentioned, the most significant variables generally are proximal. That is, they tend to be nearby even though all effects have an infinite number of material causes. There is some recognition in the popular press that infinite, rather than finite causality reflects reality. One example is the much-ballyhooed "Butterfly Effect." This is the fanciful imperiment (thought "experiment") that the flutter of a butterfly's wings in Japan might produce just enough air movement to trigger a tornado in the U.S. Nonetheless, most scientific experiments work with proximal causes to avoid transmission delays and the difficulty of measuring tiny, nearly insignificant distal causes.

Univironmental analysis and Infinite Universe Theory go hand in hand. The Infinite Universe is one, comprising the matter in motion involving the infinitely small to the infinitely large. The beauty of this way of looking at and analyzing things is its ability to answer questions that have been unanswerable throughout history. In addition, it provides the correct approach for answering the questions that will appear throughout humanity's future evolution. The next part of this book provides logical answers based on "The Ten Assumptions of Science," progressive physics, neomechanics, and univironmental determinism.

PART III: QUESTIONS RESOLVED BY INFINITE UNIVERSE THEORY

Standard practice in science, especially in physics, generally involves the presentation of an equation that can be tested. That is what we did in "Universal Cycle Theory" and in our technical papers on cycles.[112] Stephen Puetz's equation for the Universal Wave Series can be used to calculate and thereby predict natural cycles ranging from days to billions of years. There will be more equations to come, but I am sure also that they will be contested by those who use the traditional assumption of *finity*. Unfortunately, there is an inherent contradiction in the whole attempt to test **infinity** with a finite equation. As mentioned above, fundamental assumptions like these cannot be proven right or wrong. Although we will continue to gather data to support our hypothesis that the universe is infinite, we will never prove it completely. No one is ever going to the edge of the universe to answer that question. Nonetheless, Infinite Universe Theory can be used to resolve numerous paradoxes, contradictions, and questions that are unanswerable by regressive physics and cosmogony.

What follows are a series of questions I have been asked over the decades. Their answers will help elucidate what the universe might be like if it was infinite. Note that my answers are rather explicit and generally unequivocal. This is partly because I try to adhere firmly to "The Ten Assumptions of Science," strive for clarity, and detest agnosticism. The answers to these questions amount to a set of conclusions that follow from the assumptions and analyses that went before. Those conclusions form a way of

---

[112] Prokoph and Puetz, 2015, Period-Tripling and Fractal Features in Multi-Billion Year Geological Records; Puetz and Borchardt, 2011, Universal Cycle Theory; Puetz and Borchardt, 2015, Quasi-periodic fractal patterns in geomagnetic reversals, geological activity, and astronomical events; Puetz and others, 2014, Evidence of synchronous, decadal to billion year cycles in geological, genetic, and astronomical events; Puetz and others, 2016, Evaluating alternatives to the Milankovitch theory.

thinking about the universe much in the way that "The Scientific Worldview" did.

The answers to many of these questions were implied above, but I include them here for many reasons. First, the advent of the Internet has demanded short communications for short attention spans. Second, some of the questions are technical and might be of little concern to anyone who has not studied them. Third, many of these questions and answers resolve additional contradictions and paradoxes not mentioned in the main text. Again, without Infinite Universe Theory, these questions cannot be answered satisfactorily. Attempts based on regressive physics and cosmogony either are incorrect, illogical, contradictory, or just plain silly.

Chapter 13
Scientific Philosophy

The valiant efforts of Collingwood and Popper could not reach a satisfactory conclusion for lack of the overt assumption of *infinity*. Collingwood knew that philosophy rested upon assumptions that were unprovable but did not know why that was so. Popper knew that theories could be falsified and never proven but did not know why that was so. They lived when the *finity* of classical mechanics was yet to be displaced by the *infinity* of neomechanics. The Heisenberg Uncertainty Principle made *finity* untenable. But, instead of choosing neomechanics, physicists chose the indeterministic interpretations of relativity instead. Stepping backward instead of forward, physics would never accept the proposition that causality was infinite and that uncertainty was subjective. Let us begin with the answer to why you got to this point:

13.1 Does curiosity imply the advent of Infinite Universe Theory?

Yes. Curiosity involves an inquiry outside oneself—that is what you are doing by reading this book. The scientific "attitude" is based on the assumption that the truth may be known through outside activities: observation and experiment. The nonscientific attitude is the belief that truth already is known or that it may be known in ways that do not involve interacting with the external world. The scientific attitude is inherently progressive—and dangerous. Education is the awful destroyer of miseducation. You probably know about the Tower of Babel,[113] but how about this from the Qur'an:

> *O you who believe! do not ask questions about things which if made plain to you may cause you trouble... Some people*

---

[113] Genesis 11:1-9. [http://go.glennborchardt.com/Babel].

before you did ask such questions, and on that account lost their faith.[114]

It seems that folks are not too good at stifling their curiosity despite frequent admonishments to do so. In 1536, John Calvin, an early leader of the Reformation, warned "Ignorance of things which we are not able, or which it is not lawful to know, is learning, while the desire to know them is a species of madness."[115]

Pope Francis reiterates Calvin's five-century old reprimand:

And the spirit of curiosity is not a good spirit. It is the spirit of dispersion, of distancing oneself from God, the spirit of talking too much. And Jesus also tells us something interesting: this spirit of curiosity, which is worldly, leads us to confusion.
—Pope Francis

Figure 22 Pope disses curiosity. Credit: Slide courtesy of Jerry Coyne and Time.[116]

The alternative is to stifle the engine of science at an early age. Joyce Meyer leads the battle:

---

[114] Qur'an 5:101-102. [http://go.glennborchardt.com/Quran-5].
[115] Calvin, 1536, The Institutes of the Christian Religion, p. 2233.
[116] http://go.glennborchardt.com/popeandmeyer.

152 *Infinite Universe Theory*

> "DON'T REASON in the mind JUST OBEY in the spirit."
>
> "SATAN will attempt to fill your CHILD with worry, REASONING, fear, depression, and discouraging negative thoughts."
>
> "SATAN frequently steals the will of God from us due to REASONING. The Lord may direct us to do a certain thing, but if it does not make sense - if it is not logical - we may be TEMPTED to disregard it. What God leads a person to do does not always make LOGICAL sense."
>
> **DON'T REASON! JUST OBEY!**
>
> Joyce Meyer
> Battlefield of the Mind for Teens

Figure 23 Hey all you god-fearing teenagers: Don't reason! Just obey! Credit: Slide courtesy of Jerry Coyne.

The "confusion" alluded to here is an enduring problem for immaterialists who nonetheless must live in the material world. Would-be solipsists expect contact with the world to produce contradictions and paradoxes. Like those who still believe that the universe exploded out of nothing, they have learned to live with the cognitive dissonance triggered by curiosity.

The statement "Curiosity killed the cat" is not without wisdom. On the other hand, without interacting with the outside world, nothing is done. Each step, each bite of food, is an "ex"periment, a venture in the outside world. As we have seen, organized religions abhor such excursions outside their cloisters. The evolutionary purpose of religion is to instill and enforce loyalty. While this may be successful in keeping the tribe together in defense against the incursions of other tribes, the required isolation can be only temporary. In an Infinite Universe, convergence is inevitable; the incursions ultimately cannot be stopped entirely—the macrocosm is always active. If nothing else, we must learn about the external world for our own protection. Even so, there is a constant tension between learning and not learning; between stepping out with science and its preoccupation with the external world and remaining inside with its preoccupation with the internal world. The political battles rage. The proponents of education unknowingly sponsor programs and curricula designed to make us all excellent atheists and believers in Infinite Universe Theory; the proponents of religion knowingly sponsor programs and curricula designed to make us all excellent theists and believers in Big Bang Theory. This philosophical struggle between determinism and indeterminism is perpetual and progressive. We are all born not knowing much, but having learned something about the external world, we find it nearly impossible to unlearn it.

Globalization now makes it nearly impossible to maintain a cloister, sect, religion, or finite universe free from the intrusions of the macrocosm. Despite Calvin, Pope Francis, and Meyer, we are becoming ever more inclined to pursue the "species of madness," curiosity, and reason that science is based upon. In an Infinite Universe, there is no limit to that pursuit. The universe has produced matter that contemplates itself, a species that will not and cannot obey commands to avoid the tree of the knowledge of good and evil.[117] This quest is as limitless as the universe is infinite.

---

[117] Genesis 2-3. [http://go.glennborchardt.com/treeofknowledge]

13.2 Does calculus imply the advent of Infinite Universe Theory?

Yes. Those forced to study mathematics often get the impression that calculus is extremely complex. That may be the case, but there is a rather simple fundamental aspect to it as well. The abstraction we call calculus reflects the reality that the universe is infinitely subdividable and integrable at all scales. Every portion of the universe can be differentiated, that is, taken apart (dx by dx), with each successive division resulting in both matter and space ad infinitum.[118] Similarly, every portion of the universe can be integrated, that is, put together with another portion (Σdx), with each successive integration resulting in both matter and space ad infinitum. The taking apart reflects the divergence and the putting together reflects the convergence of the reality we also assume with *complementarity*.

With calculus, Newton and Leibnitz showed that, at least in the logic of mathematics, there is no "end" to infinity, as that would be a self-contradiction. Nonetheless, they did not have enough data to take what would have been the prematurely bold—and dangerous—step of proposing that the universe was infinite. At times, Newton favored *macrocosmic infinity*, but his development of mechanics clearly rejected microcosmic *infinity*. After all, that is why I developed neomechanics, which is simply classical mechanics with the assumption of *infinity*.

Here is one case in which mathematical thinking got it right. So often in regressive physics and cosmogony, we are confronted with mathematical conclusions that have no basis in reality. Indeterminists may imagine extra-Euclidean dimensions for which there is no evidence. But, even though calculus is as

---

[118] As per the Tenth Assumption of Science, interconnection (All things are interconnected, that is, between any two objects exist other objects that transmit matter and motion).

abstract as other types of math, the real world demonstrates infinite subdividability and integratability at every turn.

13.3 Does univironmental determinism imply the advent of Infinite Universe Theory?

Yes. Univironmental determinism is the universal mechanism of evolution in which what happens to a portion of the universe is determined by the infinite matter in motion within and without. Infinite Universe Theory was impossible without this realization. As mentioned, univironmental determinism also happens to be the scientific worldview. Previous forms of scientific philosophy tended to overemphasize the macrocosm (classical mechanics) or the microcosm (systems philosophy). Newton's theory of gravitational attraction, which he mentioned 185 times on the 594 pages of Principia, led him to the view that the universe was macrocosmically infinite:

> *It seems to me, that if the matter of our sun and planets and all the matter of the universe, were evenly scattered throughout all the heavens, and every particle had an innate gravity towards all the rest, and the whole of space throughout which this matter was scattered was but finite, the matter on [toward] the outside of this space would, by its gravity, tend towards all the matter on the inside, and, by consequence, fall down into the middle of the whole space, and there compose one great spherical mass. But if the matter was evenly disposed throughout an infinite space it could never convene into one mass; but some of it would convene into one mass and some into another, so as to make an infinite number of great masses, scattered at great distances from one another throughout all that infinite space.*[119]

---

[119] Newton, 1688, Letter to Dr. Covel.

13.4 What is a BS meter and why do you need one?

Everyone should have a BS (Basic Science) meter, a set of assumptions, principles, laws, or other guidelines for making judgments about statements others proclaim to be true. It should be clear by now that the BS meter I use to detect falsehoods in physics and cosmology is the set of principles that I call "The Ten Assumptions of Science." A transgression of any one of these assumptions proves a claim to be "indeterministic" and unworthy of further attention—except as a good example of unscientific thinking. As I mentioned previously, it seems that I got the idea from the University of Wisconsin, where you can see a plaque on Bascom Hall that advocates "fearless sifting and winnowing by which alone the truth can be found."[120] The BS meter helps us do that with the greatest efficiency.

Most folks have never heard of "The Ten Assumptions of Science" and have only vague thoughts about what their own principles might be. It seems that they have neither the opportunity nor the interest in confronting all the contradictions handed down while experiencing the vagaries of the education-miseducation struggle. Of course, you might not agree with some, or even all, of "The Ten Assumptions of Science." In that case, I encourage you to prepare your own list of consupponible and fundamental assumptions.

13.5 Does Infinite Universe Theory resolve the "Who Created God" question?

Yes. As children growing up in Wisconsin, my friends and I often asked this question without getting a satisfying answer. The upshot was that we came face-to-face with infinity at an early age. And so it is for anyone who contemplates the nature of

---

[120] http://go.glennborchardt.com/sift

existence. Infinity is inescapable for determinist and indeterminist alike. One look at our natural surroundings reveals infinite variety and infinite beauty. To have an answer to this question is greatly satisfying. Infinite Universe Theory teaches that nonexistence is impossible. There is no place in the universe that does not have matter in motion. The "perfectly empty space" of the indeterminist is only an idea—it cannot exist anywhere in the universe.

As mentioned, evolution is controlled by univironmental determinism (What happens to a portion of the universe is determined by the infinite matter in motion within and without). In the Infinite Universe, each microcosm is unique, forming via the convergence of supermicrocosms from the macrocosm. These hesitate for a time, with its submicrocosms eventually diverging from each other to continue throughout the macrocosm. Each microcosm has a precious, finite life, a beginning and an end. We cannot say the same for the Infinite Universe, for it is as eternal as it is necessary for the existence of the microcosms within. Newton's body, when moving through the Infinite Universe, needs no "first cause" to provide that motion. The Infinite Universe always has yet another body to provide the necessary acceleration described by the Second Law of Motion.

Chapter 14

Regressive Misconceptions

In applying our BS meter, we find that the great regression in physics and cosmology has left us with numerous mistaken beliefs. Although I have already mentioned many of these, I repeat some here in quotable essay form that can be used to confront these shibboleths whenever they rise from the regressive muck. These misconceptions are part of the regressive propaganda. They are like the "fake news" of the Internet, which adds confusion, manages to fleece the gullible, and stymies scientific progress.

14.1 Did MMX prove that aether did not exist?

No. The Michelson-Morley experiment of 1887 (MMX)[121] was a test for a fixed ether. It was an attempt to measure the 30 km/s motion of Earth relative to that supposedly fixed aether. However, because the aether around Earth is mostly entrained, just like our atmosphere, very little relative motion was detected. The experiment was akin to trying to measure the velocity of the jet stream at sea level. The MMX data were interpreted by regressive physicists as a null result, although, according to Bryant,[122] the raw data indicate that there was less than a 1% chance that a null result was obtained—it actually proved that there was entrainment. Elevation at Cleveland, Ohio where the experiment was conducted was about 200 m. Others, having used similar methods at elevations up to 1830 m were able to detect greater relative motion—indicating that aether entrainment decreases with altitude just like Earth's atmosphere.[123] My plot

---

[121] Michelson and Morley, 1887, On the relative motion of the earth and the luminiferous aether. [To shorten the many debates among reformists, this experiment is often referred to as "MMX."]

[122] Bryant, 2016, Disruptive: Rewriting the rules of physics.

[123] Borchardt, 2007, "The Scientific Worldview," figure 8-2. See also Figure 41 in the present book.

of the results yielded this equation for the velocity measured vs. altitude: $V = 183A^{1/2}$, where V = relative velocity, m/s and A = altitude, m. The relationship indicates that the full measure of Earth's velocity could not be measured with those methods at altitudes lower than the troposphere. The decrease in entrainment as a function of the square root of altitude is not just happenstance. More in Chapter 16.2.

14.2 Is the speed of light constant?

No. There are no absolute constants in nature. The speed of light is no more constant than the speed of sound. Like the speed of sound, the speed of light is dependent on nothing but the medium that transmits it. Light in "vacuum" travels at about 300,000,000 m/s, while light in water travels at about 225,000,000 m/s. Sound in air travels at 343 m/s at 20°C, while sound in water travels at 1,484 m/s. Per *relativism* and *infinity*, all these media exhibit slight variations dependent on purity, temperature, pressure, etc. Light is no different. Because neither light nor sound is a particle, the velocity of the source has no effect on the velocity of the wave. If light was a particle, its motion and the motion of its source would be additive. Clearly, they are not, as Sagnac demonstrated in 1913—a fact inducing perplexity for all those trying to make sense of the relativity mess.

In spite of that evidence, Einstein continued to insist that light was a particle. To make that supposition fit his theory, he had to give light special magical properties unknown to any other particle (Table 6). Because light is wave motion in aether, which travels at a relatively constant velocity, he gave his light particles a constant velocity too. This was strange because no other microcosms in the universe travel at constant velocities. In addition, unlike other particles, light particles had to be massless. Of course, that was necessary because light was wave motion

Table 6 Einstein's eight ad hocs.

1.	Unlike other particles, his light particle always traveled at the same velocity—it never slowed down.
2.	Unlike other particles, it attained this velocity instantaneously when emitted from a source.
3.	Unlike other particles, it would not take on the velocity of its source.
4.	Unlike other particles, it was massless.
5.	Unlike other particles, light particles did not lose motion when they collided with other light particles.
6.	Unlike other particles, any measurement indicating light speed was not constant had to be attributed to "time dilation"—another especially egregious ad hoc.
7.	Time was to be considered something other than motion, for motion cannot dilate.
8.	The claim light speed was constant flew in the face of all other measurements showing there are no constants in nature because everything is always in motion. Because the universe is infinite, every measurement of every so-called "constant" always has a plus or minus. The velocities for wave motion in any medium are dependent on the properties of that medium, which vary from place to place.

and motion has no mass—only the particles in the medium that transmits it have mass. As mentioned above, unlike other particles, light particles were not subject to Newton's Laws of Motion. Hence, the First Law of Motion did not apply because that would mean that momentum, P, would be zero if the particle's mass, m, was zero per the equation P=mv. The Second Law of Motion, F=ma, did not apply, not only because light particles were massless, but also because all light particles were supposed to travel at c. There was no way for a light particle to accelerate another light particle because both were assumed to travel at the same speed at all times.

According to the theory, this mysterious beast was the fastest particle possible. I explained this previously, so here I will try to simplify it further. The Lorentz factor, γ: $1/\sqrt{1-(\frac{v^2}{c^2})}$ shows up so much in nearly anything to do with relativity, so that it is usually designated as γ. Remember, if you accelerate some other particle with rest mass, m_o to a velocity, v, then the moving mass is supposed to be equal to γm_o. As v approaches c, γ would approach infinity, with the mass becoming infinite if the particle had any mass at all. The Lorentz factor in relativity equations is what implies that mass is a function of velocity. Now, velocity in perfectly empty space would have nothing to do with mass regardless of what the math says. To understand γ and the experiments that support it, we must be able to explain its physical cause. In neomechanics, we assume that space is not empty. Despite Einstein's early contention that aether is unnecessary for relativity, a physical explanation of the velocity/mass relationship is impossible without it. That is why regressives do not attempt that. Any microcosmic motion in the Infinite Universe must result in collisions with supermicrocosms in the macrocosm. In the neomechanics section on the absorption of motion I showed how acceleration has a tendency to increase mass. Thus, the initial mass increase observed in particle accelerators is due to acceleration. After all, that is why they are called "accelerators." Any increase in velocity requires acceleration—neomechanical collisions in which internal submicrocosms are accelerated along with the microcosm. There always will be a mass increase commensurate with the increase in velocity produced by acceleration. The speed limit for accelerators is c because the acceleration necessarily is produced by electromagnetism. The aether medium cannot transmit wave motion at velocities greater than that speed limit.

Similarly, a microcosm traveling through a macrocosm filled with aether particles must exchange motion with those particles. This is similar to what happens when the space shuttle enters the atmosphere. Collisions with the nitrogen and oxygen in the air cause the ceramic surface of the shuttle to become hot and to increase in mass due to the acceleration of the submicrocosms within. Now, γ does not include an independent factor for time. In other words, the mass attained at any particular velocity does not increase during the time that the microcosm remains at that velocity. The Lorentz factor therefore implies that, over time, the absorption of motion is equal to the emission of motion. In other words, the heating and cooling of any microcosm traveling through the aether at any velocity is equivalent.

All the fantastic contradictions above disappear when we consider light as wave motion in the aether. Again, the speed of light, like all wave motion, is dependent on the medium that transmits it. That is why light slows to 75% of its usual velocity when it travels through water. Light speed is an inverse function of the index of refraction, which is related to the concentration of baryonic matter. Thus, the index is 1.000277 for air, 1.33 for water, 1.41 for hydrogen peroxide, 1.470 for Pyrex glass, and 2.419 for diamond. By definition, the index is 1.0 for a baryonic vacuum. Again, all the media mentioned above transmit wave motion at a relatively "constant" velocity, which does not make the transmission of light particularly unique. Waves in a medium move from point A to point B at a constant rate, although the individual particles constituting the medium do not.[124] Again, unlike particles, waves do not take on the velocity of their source, which is precisely what is observed for light.

---

[124]http://go.glennborchardt.com/Lwavegif;
http://go.glennborchardt.com/Twavegif

Here is a simple demonstration on how the medium works with respect to the velocity of the source. If I dropped a rock into a lake while sitting in a boat at rest, the waves generated in the water would travel at a velocity v toward the shore. If I dropped the rock into the lake when the boat was moving at 100 mph, the waves generated would still travel at velocity v toward the shore. The motion of the boat (source) would have nothing whatsoever to do with the ability of the medium (water) to transmit the waves produced by the rock.

On the other hand, *measurements* of wave motion must consider the motion of the observer. For instance, people in another boat moving toward my boat will encounter the waves I generate much more quickly than if they were on land. The waves I generate might be 2 meters apart, and take 1 second to form each crest, but because the other boat is coming toward me, it will encounter the waves sooner and will get measurements indicating that the waves are less than 2 meters apart—if those folks neglect to account for their own motion. If they were going away from me, they would encounter the waves later and their measurements of wavelength would be somewhat greater.

The belief that light is a particle has presented numerous paradoxes for regressive physics. One is mentioned in the question concerning the application of the $E=mc^2$ equation in neomechanical interactions involving the emission of motion. Another was the fact that, unlike other particles, the imagined particle—the photon—does not take on any of the motion of the body that emits it. As an example, let us suppose that a major league pitcher threw a 100-mph fastball at you while moving toward you in a convertible at 65 mph. Disregarding friction due to air for a moment, those two velocities would be additive. You would have to be super-human to hit that 165-mph fastball! Instead, like any other wave, light travels at the velocity typical of the medium. Once again, we could drop a rock into a lake from a boat traveling at any velocity from 0 to 317 mph. The waves produced by that rock would take the same amount of time to reach the shore no matter what the velocity of the boat was. That is exactly what light does—because it is a wave in a medium. That is why aether deniers can never resolve the paradoxes involving light considered as a particle.

Of course, a wave medium always contains particles, so any contact with the medium will produce collisions we can describe with neomechanics. The photoelectric effect is an example of what happens when aether particles collide with baryonic matter. This particulate nature, however, is a property of the medium, not of the wave motion. Water without waves will still make you wet.

14.3 Does energy have mass?

No. The equation $E=mc^2$ is true nonetheless. Its interpretation has been bastardized by popularizers who often claim that energy and mass are equivalent.[125] They are not. The "equivalence" shown in this equation is simply mathematical. Energy is defined as a calculation, while mass is defined as the resistance of an object to acceleration.

The equation describes all transformations, not just those involving nuclear fission and fusion as commonly thought. As I showed in the chapter on neomechanics, the equation describes the conversion of submicrocosmic motion into supermicrocosmic motion, and vice versa. Remember that I define a submicrocosm as what is inside, and a supermicrocosm as what is outside of a microcosm (a portion of the universe). In other words, if a microcosm had no submicrocosms inside it, it would have no

---

[125] Bodanis, 2000, $E=mc^2$.

mass.[126] Now, in your study of neomechanics you learned that all microcosms must have submicrocosms ad infinitum. In review, I present Figure 24 as a simple illustration of how the $E=mc^2$ equation works. It is merely a reiteration of Newton's Second Law of Motion. One object collides with another, transferring some of its motion to the object it hits. A collision between a supermicrocosm and a submicrocosm across the microcosmic boundary causes the submicrocosm to be accelerated (Figure 24). The effect of this increase in velocity of the submicrocosm can be calculated as an increase in momentum ($P=mv$). This increase in internal motion causes the microcosm as a whole to have an increase in resistance to acceleration, which is measured as an increase in mass.

Here is a macro-example that might help in understanding that mass increases when there is an increase in internal motion:

Suppose that football team A forms a densely packed circle. Remember that mass is the resistance to acceleration. Now, suppose that football team B tries to test the resistance to acceleration of team A by trying to knock it down. Team B might be able to do it, probably by running at team A and colliding with it. Next, we ask team A to display a little internal motion, with fists and feet flying in all directions. Now, team B must run and push a little harder, because team A will be pushing back. They would be less of a "pushover." In other words, some of the force ($F=ma$) produced by the colliders of team B will be negated by the hitting and kicking ($F=ma$) produced by team A. The mass of team A will increase temporarily because its resistance to acceleration will increase temporarily. Team B must push even harder to knock team A down.

---

[126] This is another reason the photon cannot exist and cannot be part of Infinite Universe Theory.

166 *Infinite Universe Theory*

Note that the amount of "matter" attributed to team A has not changed during the experiment and that its mass will return to normal after team members stop hitting and kicking. As an aside: both teams actually will lose some mass (measured as "weight") due to the exercise. The motion we measure as calories "burnt" in the process will be emitted to the atmosphere and the aetherosphere in the form of heat and radiation as shown in Figure 24. The temperature of the air surrounding the exhausted athletes will rise, and infrared radiation will tickle the aether (Figure 25).

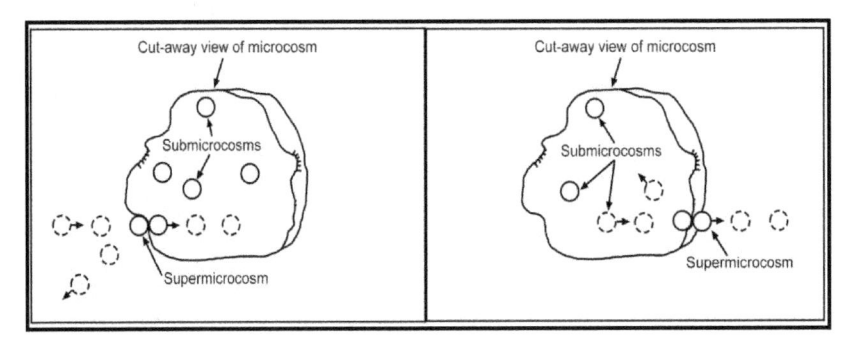

Figure 24 Neomechanical interactions apropos the E=mc2 equation illustrating that both absorption and emission involve mechanical collisions described by Newton's Second Law of Motion (from Figure 16 and Figure 18). By denying aether exists, regressive physics denies that these collisions occur.

Figure 25 Infrared photo of radiation from chimps. Credit: Cool Cosmos/IPAC, NASA/JPL-Caltech.[127]

While the total matter and motion in the universe remains constant per *conservation*, the mass of each microcosm is either constantly increasing or decreasing. This is because all submicrocosms within microcosms are amenable to impacts from supermicrocosms. These are, in turn, amenable to contributing some of their submicrocosmic motion across the microcosmic boundary to supermicrocosms in the macrocosm (Figure 24). This transfer of motion to the outside is inexplicable without some kind of supermicrocosm that can receive it. Here is the kicker: for aether deniers there often is nothing but perfectly empty space to receive this motion. They are forced to imagine a kind of matterless motion that takes on the ghostly form called energy even though energy neither exists nor occurs.

Mass/energy transformations are simple once you break them down to interactions involving matter in motion. What makes the

---

[127] http://go.glennborchardt.com/coolcoscaltech

understanding so difficult is the required change in philosophy in which one needs to relinquish the erroneous interpretation of the energy concept and adopt the aether concept instead. Note also that the $E=mc^2$ equation must work at all levels, in the same way as neomechanics. All microcosms, no matter how large or small, must contain submicrocosms ad infinitum. The finite particle hypothesized by indeterminists is "solid matter." It has no submicrocosms by which the equation could possibly work. That is yet another reason Finite Particle Theory is a contradiction of Infinite Universe Theory.

You may have heard that Einstein's equations claim that mass depends on velocity. However, there would be no physical reason for mass to increase when velocity remains constant during inertial travel through perfectly empty space. Any actual mass increase could occur only because space is not perfectly empty. As seen above, acceleration, which is required to increase velocity, always requires a collision from a faster microcosm. The presence of aether changes the situation. At "constant" velocity, collisions with aether particles would tend to cause deceleration and a temporary temperature and mass increase similar to what occurs when a space capsule enters the atmosphere. Because aether particles are so tiny, these effects normally are insignificant.

Let me repeat: The collision of a supermicrocosm with a microcosm always produces some acceleration of submicrocosms within the microcosm, which is realized as heat and an increase in mass (Figure 24). As shown in the neomechanics chapter, the transfer of motion from hammer to nail is what increases its internal submicrocosmic motion and increases its mass. Cooling reverses the process: "As the coffee cools, mass is lost."[128]

Again, by erroneously rejecting the aether, Einstein imagined that a collidee was unnecessary. Of course, this produced a huge step backwards, and caused all sorts of theoretical havoc. Ever since, relativists were not able to accept the theoretical necessity for aether despite Sagnac's proof and de Sitter's confirmation. On the contrary, as indeterminists, trying to avoid the implications of *infinity*, they are more likely to imagine stuff more fantastic than aether, such as massless particles, matterless motion, and immaterial fields.

14.4 Does "dark energy" exist?

No. Energy, whether it is called light or dark, does not exist. Energy is a calculation we use to describe matter in motion and the motion of matter. The hypothesized "dark matter" of the astronomers must exist and, of course, must be in motion per *inseparability*. As a matter-motion term, energy has been widely considered motion by some and matter by others. This confusion reigns supreme as regressive physicists and cosmogonists, by definition, do not know what energy is. The official view, reflected in this rash statement from NASA, implies that energy exists and that it is a material constituent:

> ...roughly 68% of the Universe is dark energy. Dark matter makes up about 27%. The rest—everything on Earth, everything ever observed with all of our instruments, all normal matter—adds up to less than 5% of the Universe.[129]

NASA's explanation amounts to a vision of bare-naked matterless motion floating around outer space—an impossibility, although that view appears necessary for regressive physics.

---

[128] Gardner, 1962, Relativity for the million.
[129] NASA, 2014, Dark Energy, Dark Matter.

Dark energy is merely an ad hoc invention needed to keep the Big Bang Theory alive. Once Einstein's Untired Light Theory is rejected in favor of the aether concept, the expanding universe will be history. No longer will cosmologists need "dark energy" to propel the imagined expansion.

Is there dark matter? Of course. There is no reason for all matter to be "light," that is, luminous. In "Universal Cycle Theory," we speculated that much dark matter exists in the universe as undetected planets without luminous central stars. Whatever it is, dark matter must be matter in motion. Could we calculate its energy? Sure, in the same way we can calculate energy for any other microcosm in motion. Could dark energy exist independently of dark matter? No. Could dark energy be a separate constituent of the universe? No. Without matter, the calculation $E=mc^2$ would be $E=0$ mass times c^2, which would equal 0.

I suspect much of the confusion about dark energy stems from aether denial. Aether pervades the universe, with aether particles being tiny, but nonetheless having mass just like everything else. Thus, energy calculations involving so many aether particles would result in huge values. Now, dark matter was initially proposed to explain the tendency for rotating spiral galaxies to appear more massive than non-rotating elliptical galaxies despite their having equivalent luminous matter. In "Universal Cycle Theory," we speculated this was a result of vortex formation in which particle size and density follows Stokes's Law in an aether medium. This law describes the sedimentation rates for particles in fluvial and lacustrine environments on Earth, in which gravels are deposited first, followed by, sand, silt, and clay. Cosmic vortices do the same thing, with the rotating Earth having increasing density with nearness to its center and with the Sun being 99% of the mass of the solar system. As vortices age, gravitation pushes increasing amounts of matter toward the center. Thus, the Milky Way is relatively young, with most of its mass outside the nucleus. In particular, nonluminous planet-size bodies on the periphery of the galaxy would not be observed by our telescopes, although they would contribute to its mass.

14.5 Does the "god particle" exist?

No. Many regressive physicists hypothesize an irreducible (elementary particle) or "god particle" in what amounts to a continuation of the moribund Finite Particle Theory. The idea is based on the assumption of microcosmic finity, which was the primary supposition underlying Greek atomism and its offspring, classical mechanics and classical determinism. As mentioned, the Heisenberg's Uncertainty Principle implied *finity* was untenable. The adoption of ***infinity*** makes neomechanics a viable replacement for relativity. Microcosmic infinity, in particular, is a thread running throughout all ten assumptions of science. It is what makes them "consuponible," that is, if you can assume one of them, you must be able to assume all of them without contradiction. For instance, the Second Assumption of Science, ***causality*** (All effects have an infinite number of material causes), would not be possible without it. All things (microcosms) must be bathed in an infinite sea of particles (supermicrocosms) so that no two reactions can be exactly alike. Similarly, ***interconnection*** obviously requires it.

Despite their claims to be "relativists," modern physicists cannot disentangled themselves from *finity*, the one presupposition that most distinguishes classical mechanics. They are ambivalent about the macrocosmic end of the universal hierarchy, as shown by the currently popular oxymoronic "parallel universe" and "multiverse" theories. Most are firm believers in microcosmic finity despite all the smashed atomic particles and new subparticles produced with each improvement in accelerator design. Like every "proof" of relativity, the billion-dollar bet at CERN was sure to discover another entry for the microcosmic finity hall of fame. The latest candidate to be so anointed is called the Higgs boson, a hypothesized elementary particle otherwise dubbed the "god particle" by the popular press. There is only one problem: it cannot possibly exist.

There are many reasons for this. One is its theoretical association with the photon, the oxymoronic massless light particle hypothesized by Einstein. Another is our assumption all microcosms must contain submicrocosms ad infinitum. As mentioned, an elementary particle has no parts. If it did, then *these* would need to be still more elementary. Despite the Nobel-pandering kerfuffle and its being nominated for Time's "Person of the Year" (Figure 26), the Higgs Boson is a highly unlikely candidate for the particle giving everything mass. It supposedly decays into four muons within 10^{-22} seconds. Looks like this little bit of accelerator rubble not only giveth mass, but also taketh it away pretty quickly. It has 125 times the mass of a proton and 250,000 times the mass of an electron. Admittedly, it does not exist inside any particles, but exists outside of them—sort of like the forbidden aether. The idea that it forms a kind of "molasses" in empty space to give particles mass is only a bit more than a *non sequitur*. At least it gave aether deniers a chance to save face. Looks like at least some of them are starting to agree "space" is not so empty after all.

Actually, like much regressive theory, the "molasses" idea is not without at least a tiny speck of merit. Remember mass is resistance to acceleration. Normally, we use a colliding microcosm of known mass and velocity to accelerate a microcosm and thus determine its mass. We usually assume that the environment of that microcosm has no significant supermicrocosms that would interfere with the measurement. But what if it does, as hypothesized by Higgs? Any "molasses" in the environment would need to be pushed aside during the measurements. We would require an increase in force in the same way it would take more force to push our car out of deep snow than it would if there was no snow at all. If we wanted a correct measurement of the mass of a microcosm, we would need to subtract the effect of the resistance of its macrocosm. In this case, we would not say the "molasses" or the snow gave mass to the microcosm of the car.

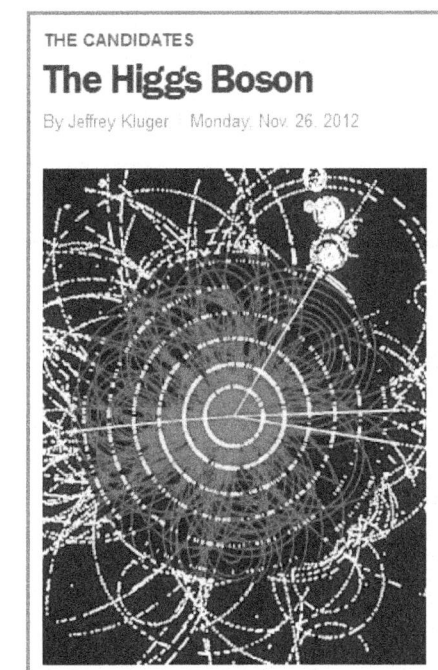

> THE CANDIDATES
>
> ## The Higgs Boson
>
> By Jeffrey Kluger   Monday, Nov. 26, 2012
>
> **What do you think?**
>
> Should The Higgs Boson be TIME's Person of the Year 2012?
>
> 19.74% Definitely
>
> 80.26% No Way
>
> Simulation of a Higgs-Boson decaying into four muons, CERN, 1990.
>
> SSPL/Getty Images
>
> Take a moment to thank this little particle for all the work it does, because without it, you'd be just inchoate energy without so much as a bit of mass. What's more, the same would be true for the entire universe. It was in the 1960s that Scottish physicist Peter Higgs first posited the existence of a particle that causes energy to make the jump to matter. But it was not until last summer that a team of researchers at Europe's Large Hadron Collider — Rolf Heuer, Joseph Incandela and Fabiola Gianotti — at last sealed the deal and in so doing finally fully confirmed Einstein's general theory of relativity. The Higgs — as particles do — immediately decayed to more-fundamental particles, but the scientists would surely be happy to collect any honors or awards in its stead.

Figure 26 Higgs-Boson propaganda proffered by Time.[130]

---

[130] Kluger, 2012, The Candidates: The Higgs Boson.

Now for the "speck of merit" that might possibly produce an increase in mass within our microcosm. If you followed previous discussions carefully, you already know how this can occur. Any collisions produced by supermicrocosms cause submicrocosms within a microcosm to be accelerated (Figure 24). Having an increase in momenta, these submicrocosms impact the microcosmic boundary, producing an increase in resistance to acceleration (i.e., "mass"). The same process occurs when a microcosm collides with a supermicrocosm absorbing motion via Newton's Third Law of Motion as well as losing motion to the supermicrocosm via Newton's Second Law of Motion.

The space shuttle is a good example of mass increase via these microcosmic-macrocosmic (univironmental) reactions. Upon encountering Earth's atmosphere during the return trip, the shuttle absorbs motion from collisions with nitrogen and oxygen molecules in the air. Consequently, the heat shield gets hot, with this increase in submicrocosmic motion appearing as an increase in mass. So I suppose one could say that the "molasses of the atmosphere gave the shuttle mass." Of course, that mass would disappear on the ground during the cool-down phase, with the increased internal motion being emitted as heat (electro-magnetic radiation) motion per the $E=mc^2$ equation. (Figure 24).

The Time article is another illustration of how the popular press glorifies Einstein and regressive physics while getting the physics all mixed up. In his paean to the Higgs Boson Jeffrey Kluger writes:

> *Take a moment to thank this little particle for all the work it does, because without it, you'd be just inchoate energy without so much as a bit of mass. What's more, the same would be true for the entire universe. It was in the 1960s that Scottish physicist Peter Higgs first posited the existence*

of a particle that causes energy to make the jump to matter.[131]

Maybe Jeffrey should review my E=mc² paper.[132] It's not complicated, and he might learn something. You will remember energy is defined as a calculation. Energy neither exists, nor occurs. There is no such thing as "inchoate energy"—only inchoate ideas about the nature of energy. The first and second sentences of the quote amount to a gross violation of ***inseparability***. Of course, a reporter might be excused, what with all the talk by regressive physicists about "dark energy" as though it was indeed a "thing." As with a lot of mainstream woo-woo press, this one claims once again that Einstein was right. Kluger says the Higgs Boson "finally fully confirmed Einstein's general theory of relativity." This is a gross distortion. The Higgs has nothing whatsoever to do with General Relativity Theory. And even if it did, there is no way any theory could be "finally and fully confirmed"—evidence only supplies support for a theory. Because the universe is infinite, theories cannot be proven true, although they certainly can be proven false. At best, the Higgs boson—if it actually existed—would prove Einstein wrong. If the boson was the "external syrup responsible for giving mass to all baryonic matter," it would prove space is not perfectly empty as claimed by Einstein and other aether deniers. In particular, it would falsify the usual indeterministic interpretation of Special Relativity Theory, with its imagined conversion of mass into pure energy that flits through empty space like the equally imaginary "soul" on its way to heaven.

14.6 Is string theory valid?

No. Any theory hypothesizing more than three dimensions has no bearing on the real world. None of the predictions

---

[131] Ibid.
[132] Borchardt, 2009, The physical meaning of E=mc².

attributed to such mathematical games has yielded any valid results.[133]

14.7 Does space-time exist?

No. As we saw in Chapter 8, without the presumption of space-time, the Big Bang Theory and its expanding universe hypothesis could not have survived the first nervous laugh. That is why Einstein's General Relativity Theory, which required space-time, must sink or swim along with the Big Bang Theory. Space-time, however, is a matter-motion term. As such, space-time neither exists nor occurs. Its objectification, prohibited by neomechanics, is what keeps the Big Bang Theory alive.

The weirdness of relativity and the Big Bang Theory does not necessarily involve the equations or the data used to defend it, but in the interpretations of those items. After all, skillful, sophisticated folks working within the paradigm probably can ignore the contradictions I pointed out in previous chapters. What you are developing, I hope, is an appreciation that there is more than one way of looking at things and that outright contradiction within a theory is not to be tolerated. Even so, what I will ask you to do is so philosophically radical it will take awhile. After all, it took Stephen Puetz, my coauthor on "Universal Cycle Theory" about three months to realize energy did not exist—and he is one of the smartest folks I know.

Now I will try to convince you that space-time is a matter-motion term. This is tough because it requires another philosophical leap for which few are prepared in these days of rampant relativity. You would think it would be the other way around—everything we know has three dimensions, why not the universe? The propaganda promoting Einstein's "genius" has

---

[133] Smolin, 2006, The Trouble with Physics: The Rise of String Theory, the Fall of a Science, and What Comes Next.

gotten most of us believing in the 4-D phantasmagoria. Smart folks actually think space-time exists. They even try to present models of it to a skeptical lay public. Nevertheless, like the other matter-motion terms, space-time is only an idea. The math may be elegant, but that does not make the universe 4-dimensional. To explain this fully, I need to clarify what I mean by both space and time.

Space

The philosophical confusion begins with the definition of space. Here again, critical choices must be made. Either space is something, or it is nothing. It is the job of all indeterminists to claim, as positivists do, that space is nothing. After all, one can do the math without having to decide whether space is material or immaterial, as Einstein and many physicists have shown. If space really was perfectly empty, it would be a boon to immaterialists everywhere. Along with data showing that space is not empty, neomechanics provides the answer with *infinity* and *interconnection*. These imply, as Aristotle assumed long ago, that the universe is infinitely subdividable. In other words, not only is the universe infinite externally, it is infinite internally as well. We observe the large and the small, with little to distinguish them except relative size. There is no reason to believe we should be able to see the largest thing or the smallest thing in the universe. Existence is obvious and there is every reason to believe existence is the primary characteristic of every portion of the Infinite Universe. These assumptions predict that one thing the Infinite Universe cannot do is to produce perfectly empty space.

Indeed, looking at the real world, we can find no examples of perfectly empty space anywhere. Thus, in the laboratory, we are unable to make a perfect vacuum. Even outer space is not perfectly empty as Einstein first assumed. Cosmic Microwave

Background data show intergalactic space has a temperature of 2.7°K (Figure 27). Temperature is the vibration of matter, so there has to be matter there, whether one calls it "aether," "dark matter," hydrogen, or whatever. Decades ago, one wag called intergalactic space a particle zoo.[134] When you look into it, you will find that any 3-D portion of the universe, no matter how small or how large, will always contain matter. Again, it is merely a question of relative scale.

There is one other way to consider what seems to be the material-immaterial nature of space. All things in the universe contain what we perceive to be matter and space, per the matter-space continuum. I write, "perceive to be," because matter is an abstraction. There is no such thing as "matter" *per se*.

Figure 27 Cosmic microwave background radiation measured by the Wilkinson Microwave Anisotropy Probe (WMAP) showing the heterogeneous/homogeneous nature of intergalactic temperature. The light areas are greater than 2.7°K and the dark areas are less than 2.7°K, although the variation is tiny: 5 X 10^{-5} °K.

---

[134] Industrial Research (a somewhat unconventional free journal that influenced me in the '70s). Greiner and Hamilton, 1980, Is the vacuum really empty?

Time

Time is motion. That statement appears as a shock to most folks. It is a mainstream understanding that time is supposed to be a mystery. Ask around. You will get numerous contradicting answers. In particular, few, if any, regressive, remunerated physicists will know what time is. The consensus seems to be that time is a dimension or a measurement (as though the dinosaurs did not experience time because they could not measure it). Like the determinism vs. free will debate, the philosophical struggle over the nature of time is endless. Folks entertain all sorts of fantasies with regard to time. Of course, in the real world, time, like all phenomena, must fit one of two categories: either matter or the motion of matter. Not being "part" of the universe, a "piece" of time cannot be examined as we do with material objects. Time is not a thing, but the motion of things. Not being a thing, time cannot shrink or swell. This is why we say time has no existence. Unlike material things, time does not have three dimensions and location with respect to other things. Universal time is the motion of all things with respect to all other things. While it is impossible to measure universal time, we can measure specific time with regard to the relative motions of specific portions of the universe (e.g., rotation of Earth, the flow of sand in an hourglass, and all sorts of clocks). Again, because time does not exist, it has no dimensions. People often confuse time with its measurement. We cannot measure time without using 3-dimensional objects that move at least in one direction in relation to other 3-dimensional objects. Nevertheless, that does not make time an object. The objects measured stay objects and the motions measured occur as motions, not as objects. Thus duration, such as a minute or an hour, is not a thing. It is not an xyz portion of the universe although there is a tendency for us to imagine it so.

Dimensions

There are only three dimensions, as anyone with a sense of touch can prove just by reaching out to another object. All objects, things, or microcosms have what we define as three dimensions. However, mathematical imagination knows no such limitation. Today's common belief in extra-Euclidean dimensions is mostly due to Einstein's objectification of motion.[135] In Special Relativity Theory we see it in his assumption that light and time are matter instead of the motion of matter. In General Relativity Theory we see it when he objectifies time as an extra dimension that eventually gave succor to the universal expansion hypothesis.

Why would theoreticians ever consider time to be a dimension? They would simply because they do not believe in *inseparability*. Its indeterministic opposite, *separability*, proposes matter is one thing and matterless motion is still another. Math does not prevent one from considering time as matterless motion, some invisible thing that flows by, never to return. I frequently plot time in my own analyses, but I have not concluded time is a thing. Nonetheless, the act of plotting time on a piece of paper is akin to objectifying it. Still, one must never fall into the trap of defining time as a dimension.

Theories dealing with more than three dimensions are simply mathematical hocus-pocus. They may be nice mental games, but they cannot apply to reality. It is true they do serve as wonderful illustrations of the kind of stuff one can get published just by adhering to the indeterministic assumptions of the funders. Some of the sharper mainstream physicists, such as Lee Smolin, are not impressed.[136] It always amazes me that ideas involving more

---

[135] Borchardt, 2011, Einstein's most important philosophical error.
[136] Smolin, 2006, The Trouble with Physics.

than three dimensions are taken seriously. Note this from Wikipedia,[137] ever the cheerleader for regressive physics and fundable mainstream inanities:

> One notable feature of string theories is that these theories require extra dimensions of spacetime for their mathematical consistency. In bosonic string theory, spacetime is 26-dimensional, while in superstring theory it is 10-dimensional, and in M-theory it is 11-dimensional. In order to describe real physical phenomena using string theory, one must therefore imagine scenarios in which these extra dimensions would not be observed in experiments.

In other words, string theory could never meet the criteria for a scientific theory: being amenable to observation and experiment.

Those who claim dimensionality for time clearly are on the indeterministic side of the philosophical struggle. One can plot time in the 3-D world of course, but that does not make it a dimension. If you now can accept the idea time is motion; we can proceed to the real meaning of "space-time."

Space-time

For those now convinced space is matter and time is motion, we can continue with the analysis of space-time, which I claim to be a matter-motion term like momentum, force, and energy. Like all matter-motion terms, space-time is an idea, concept, or calculation we use to describe aspects of the universe. Again, like the other matter-motion terms, space-time neither exists nor occurs. Our 3-D universe does not "consist" of space-time, so all the attempts to produce a physical model of it are completely futile. They amount to whimsical contributions to the determinism-indeterminism philosophical struggle. As mentioned, the idea space-time is real rather than imagined was

---

[137] http://go.glennborchardt.com/string accessed on 20161125.

critical for extending the life of the priest's Big Bang Theory. Otherwise, our being at the center of two trillion galaxies would make no sense except perhaps to those who still thought the whole thing was created just for us.

Nonetheless, the cosmogonists still have most people believing the universe exploded out of nothing—a grand *creation* that puzzles even Hawking when he ponders "what came before." A thoroughly indoctrinated physics professor once told me I had no existence, but the occasion of my birth did! Next to this, the virgin birth of Christ is the height of scientific rationality.

What then is space-time, and how should we use it? As I mentioned, space-time is a concept or idea. Thus, yesterday I sat at my desk, occupying a certain 3-D space. Today, I occupy the same 3-D space. The two material spaces are essentially the same, but all things in the universe have kept moving in the meantime. I can imagine myself sitting at my desk yesterday and I could expound on how different my "space-time position" was yesterday as opposed to today, what with the rotation of Earth, etc. Of course, none of those space-time positions exists despite the wild claims and the mathematical mixing of space and time. What exists is space and what occurs is time. Space-time may be a good visualization, but it does not exist.

14.8 Is matter a result of quantum fluctuations?

No. Baryonic matter forms from aether particles (see Chapter 16.4). As aether deniers, however, Big Bang cosmogonists labor to explain how the entire universe exploded from nothing. To see how ridiculous it can get, see the interview[138] with "renowned cosmologist" Lawrence Krauss, a professor at Arizona State who was pitching his New York Times best seller, "A universe from

---

[138] http://go.glennborchardt.com/TSWBBA

nothing."[139] As Rick Dutkiewicz implied in our PSI review,[140] Krauss really does not know where matter comes from. Still, he is excellent at emphasizing long-standing indeterministic concepts invented by aether deniers who thus had no other choice. An important one is the idea of "virtual particles." Of course, in the rational world there are no virtual particles. Either particles are particles or they are not. Particles, like all microcosms, take up xyz space; they do not blink in and out of existence as Krauss imagines. Per *conservation*, matter is eternal, whether in the form of particles, or anything else. The most that can happen is for matter of one type to be transformed into matter of another type. As mentioned, I speculate that baryonic particles form from aether particles. The transformation can occur in reverse, so the word virtual (Webster: "very close to being something without actually being it") was required if "aether" was to be dismissed as "nothing."

In the interview, Krauss said, "I think it is virtually certain that everything we see came from empty space…all the physics I know is highly suggestive that our universe popped into existence as a quantum fluctuation." These imagined magical occurrences also were repeated regularly by Hawking and other prominent cosmogonic propagandists as if they deserved serious consideration. The old "quantum fluctuation" argument is a blatant violation of the First Law of Thermodynamics, *conservation*. You cannot make something out of nothing no matter how fancy your language.

Here are some more outlandish quotes from the interview:

> *if... quantum mechanics was applied to gravity, space and time would have become dynamical and so would have*

---

[139] Krauss, 2012, A universe from nothing: Why there is something rather than nothing.
[140] http://go.glennborchardt.com/Doogie

spontaneously appeared. So you wouldn't have needed pre-existing space. Instead the space itself would have arisen.

If you wait long enough, no matter how small the probability is, it must arise. If you have particle pairs with a gravitational attraction that is just right for their total energy to be zero, you're guaranteed that something will arise from nothing.

If you can believe all that, there is a bridge I can sell you…cheap.

14.9 Does the double-slit experiment prove light is both a wave and a particle?

No. Light is a wave. Indeterminists, wanting to believe Einstein's particle theory of light, commonly have trouble distinguishing between particles (matter, which exists and has xyz dimensions) and waves (motion, which occurs but does not exist). For motion to be transmitted as a wave, it must occur in a medium filled with particles. Like the folks who produce the "wave" in the stadium, none of the microcosms actually goes anywhere. The microcosms in a medium go back-and-forth, up-and-down, or side-to-side. After the wave has passed, each microcosm assumes its original position, in the same way all attendees remain at their seats.

All waves in media containing relatively unattached particles display the phenomena seen in the double-slit experiment. This includes water and air as well as light. When motion within a medium is transmitted through a single slit, it tends to spread out evenly in all directions from the source (Figure 12). Each wave impacts the nearest barrier at about the same place. However, when the motion is transmitted through two slits, the two waves interact with each other (Figure 28). When transmitted through one slit, they do not (Figure 29).

Glenn Borchardt 185

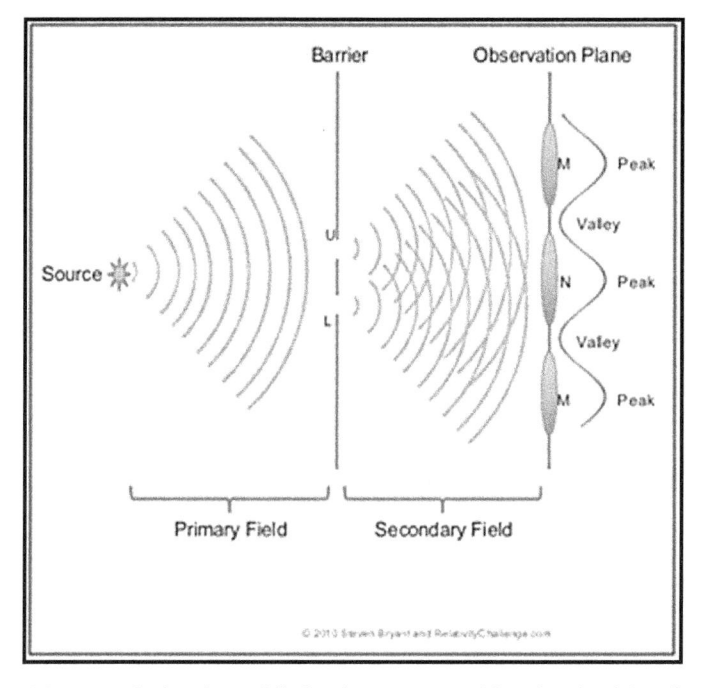

Figure 28 Wave behavior of light demonstrated by the double slit experiment (Credit: Steven B. Bryant).

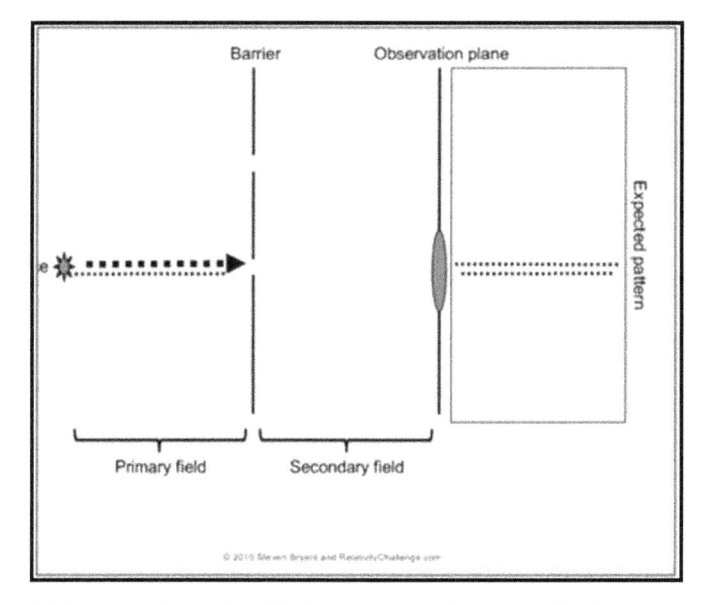

Figure 29 Expected result if light was a particle (Credit: Steven B. Bryant).

When the crests of waves coming from different directions collide, they form a single crest combining the amplitudes of both. This is "constructive interference." When the valleys of two waves collide, they form a single valley combining the amplitudes of both. Again, this is "constructive interference." When a crest and valley collide, the height of the crest and the depth of the valley combine to cancel each other out. This is "destructive interference," which results in no wave at all. Alternating constructive and destructive interference can be seen as a series of bands wherever waves collide with the nearest barrier.

If light was a particle, there would be no bands. Instead, the hypothesized "particles of light" would behave as bullets shot from the nearest slit (Figure 29). They would only collide with the nearest barrier directly across from the slit. Now, the bands we observe actually are effects produced by particles too, but these are only from particles in the last wave in the chain reaction. It is as if you accidentally got slapped in the face by the last guy to perform the wave at the stadium. Like that guy, the last particle in the chain of light motion will return to its former position, although only after having lost some velocity due to its collision with the barrier. Remember, no effect of any sort ever occurs without a collision of something with another thing.

Light has been known to be a wave ever since 1678 when Christian Huygens first publically used that interpretation in opposition to Newton's corpuscular theory of light.[141] The double-slit experiment presumably was first performed by Thomas Young in 1801.[142] I have often wondered why regressive

---

[141] Huygens, 1690, Treatise on light.
[142] Young, 1804, The Bakerian Lecture.

physicists have trouble understanding such a simple experiment. Perhaps it is because they fail to see the difference between the short-range particle motion occurring in media and the long-range particle motion occurring in spite of it. Then again, if you are in aether denial as most are, the production of interference bands must remain a mystery. Without a medium encompassing your double slit apparatus, there would be no rational way to explain its production of the interference pattern. You might even have to invent an ad hoc such as wave-particle duality, in which your erstwhile particle travels through perfectly empty space, bringing its own waves along with it!

14.10 Does Infinite Universe Theory mean everything is possible?

No. There are an infinite number of impossibilities, as well as an infinite number of possibilities in the same way there are an infinite number of even numbers and an infinite number of odd numbers. To distinguish between impossibility and possibility, we must rely on our BS meter, "The Ten Assumptions of Science." For example, it is impossible for there to be two identical things—there never will be another *you* in the universe. That is forbidden by the Ninth Assumption of Science, ***relativism***. Another impossibility is the explosion of something from nothing, as in Big Bang Theory. That is forbidden by ***conservation***. It will never happen, no matter how long Krauss waits for his magical probability to do the job.

All possibilities depend on their univironments, the interactions between microcosms and their macrocosms. While no two microcosms can be identical, they can be similar. Another way of looking at it is this: To be identical, any two microcosms must occupy the exact same space—an impossibility. That is why we observe trillions of snowflakes, with all of them forming under similar, but not identical conditions, starting with a microcosm containing water and a macrocosm at freezing temperature. No wonder no two snowflakes are alike, with the average snowflake containing about 10^{19} water molecules. [143] Obviously, one impossibility would be for a snowflake to form at room temperature.

As implied above, with respect to possibilities and impossibilities, you must be careful with the use of probability. *Uncertainty* treats probability theory as an attempt to use mathematics to measure what know and what we do not know. Probability, like Infinite Universe Theory in general, does not mean "everything is possible." For instance, both humans and electrons are variable, with no two of them having the same mass. There is a distribution, usually described by a bell-shaped curve, a useful mathematical construct. However, like other mathematical idealizations it has limitations when applied to the real world. It does not mean, for instance, that there could be either a 10,000 lb. human or a 10 lb. electron even though probability theory does not rule that out.

14.11 Is the "Twin Paradox" Resolvable without Relativity?

Yes. The "twin paradox" has been repeated and debated by regressives and reformists alike ad nauseam. It goes like this: "the twin paradox is a thought experiment in special relativity involving identical twins, one of whom makes a journey into space in a high-speed rocket and returns home to find that the twin who remained on Earth has aged more."[144] Because all motion is relative, the stay-at-home twin thinks just the opposite, that the travelling twin has aged more. Paradoxes, of course, always are founded on incorrect assumptions.[145] According to

---

[143] Roach, 2007, "No Two Snowflakes the Same" Likely True.
[144] http://go.glennborchardt.com/twinparadox
[145] Borchardt, 2008, Resolution of the SLT-order paradox.

Steven Bryant, the resolution simply involves averaging the two Doppler shifts observed by the travelling twin and the Earth-bound twin.[146] By using a train example, Bryant nicely describes what actually is happening. He imagines two equally solipsistic observers, each with frequency counters. That is, each measures time from his own perspective. In Case 1, both agree on the frequency when the train is not moving. But:

> *"Now consider Case 2, where the train is moving away from the station at velocity v. Person A, at the train station, blows his horn, which is picked up on both frequency counters. However, due to the Doppler shift and resulting lower frequency observed on the train, the frequency counter on the train is incrementing more slowly than the frequency counter at the station. In other words, the frequency counter on the train will take longer to reach 1000 beats than its counterpart at the station: The clock on the Train is running slower.*
>
> *As a continuation of Case 2, consider the situation where Person B, who is on the train, blows his horn. The sound is picked up on both frequency counters. In this case, again due to the Doppler shift, the frequency counter at the station will run slower than its counterpart on the train. Once again, a wavelength based model explains this using Doppler shifts, while Einstein's length based theory explains it as time dilation. However, the length based explanation also creates a new artifact, or paradox: From each person's perspective, his clock is correct and it is the other person's clock that is running slowly"*[147]

The twin paradox goes to the heart of relativity and the derivation of the Lorentz factor itself. Signals travelling through a medium take time. If there was no aether medium and if light traveled at infinite velocity, γ would be 1.0 and no paradox and no "time dilation" could be proposed. Einstein's solipsism and

---

[146] Bryant, 2011, The twin paradox: Why it is required by relativity.
[147] Ibid, p. 3.

positivism is also why he came out against simultaneity. Positivists and their cousins, the operationalists, are extremists descended from the empirical tradition in the philosophy of science. They might say: "If I cannot sense it or measure it, it does not exist or occur." Now, light from the Sun takes time to reach us. If the Sun ever disappeared, we would not know about it for 8 minutes. Any measurement we could make would not prove that the Sun and Earth existed at the same exact time. Nonetheless, most normal folks "assume" that the Sun and the Earth exist simultaneously even though the proof is not possible. That was the lesson we learned from "The Ten Assumptions of Science" as well. Assumptions are required just to live our lives. The reluctance of regressives to support that view only betrays the fact their hidden presuppositions are suspect.

Chapter 15

Tests of Relativity

Unlike well-accepted theories, their claims defying common sense have always made Special Relativity Theory and General Relativity Theory a matter of contention. Relativity seems to have been both confirmed and falsified many times. There are several reasons for this confusion.

First, the equation $E=mc^2$ Einstein appropriated from Maxwell, is correct, has much experimental support, and is a critical part of Infinite Universe Theory as well. However, it was not derived by Einstein, and when interpreted properly, really has nothing to do with the illogical claims of relativity.[148] Considering it so would be like considering Newton's Second Law of Motion as providing experimental support for relativity every time two billiard balls collide.

Second, tests of γ, that is, $1/\sqrt{1-(\frac{v^2}{c^2})}$, which was used throughout Special Relativity Theory usually involve acceleration, not velocity *per se*. Simply traveling at a high velocity through perfectly empty space per Newton's First Law of Motion would not really produce time gains or cause mass to increase. Nevertheless, as shown in the neomechanical explanation of $E=mc^2$, acceleration always increases mass and deceleration decreases it. This is because a microcosm cannot be accelerated without a coincidental increase in the motions of its contained submicrocosms and *vice versa* for deceleration. In a vacuum lacking aether particles, the mass of a 1-kg microcosm

---

[148] For instance, the actual equation, $E=mc^2$, was not included in his famous 1905 papers introducing Special Relativity Theory: Einstein, 1905, On the electrodynamics of moving bodies; Einstein, 1905, Concerning an Heuristic Point of View Toward the Emission and Transformation of Light.

traveling at high velocity is the same as the mass of a 1-kg microcosm traveling at low velocity. From the neomechanical point of view, the only way velocity could affect mass would be for the microcosm to absorb motion from the macrocosm. In other words, in lieu of any other supermicrocosms, aether particles would have to be present. Changes in velocity (acceleration) require a push from supermicrocosms. When the accelerations are tiny (as in gravitation), increases in mass also will be tiny and the unidirectional kinetic energy equation $KE=1/2\ mv^2$ will apply.[149] Because the submicrocosms are sped up, this internal motion may appear as heat, which subsequently may be emitted bi-directionally as radiation per the $E=mc^2$ equation. Large mass increases will be noticed only when the colliders have velocities near c. Similarly, that is what happens when the microcosm becomes the collider, crashing into supermicrocosms, absorbing some of their motion, gaining mass in the process. In both instances, the microcosm will resume its former mass when it is decelerated to its former low velocity.

Third, the most common falsifications of relativity generally occur only when time dilation must be proclaimed to protect the sanctity of c or the assumption aether does not exist or that light is a special particle. These falsifications go to the guts of relativity and its ad hoc assumptions, while the two types of confirmations mentioned above adhere to neomechanics. Objections to the falsifications generally involve devotion to indeterministic assumptions and the corresponding ignorance about the nature of light and time.

Fourth, relativity requires the objectification of time, which is the philosophical error that will always bedevil Einstein's

---

[149] Speculation: Earth's core presumably is hot because of radioactive decay. However, is it not possible that some of the core's submicrocosmic heat has been produced during geologic time by the aether impacts due to gravitation?

reputation.[150] In the previous discussion, it became clear that without General Relativity and its inclusion of time as a dimension, space-time could not have saved the expanding universe interpretation necessary for the Big Bang Theory.

15.1 Did Sagnac prove the existence of aether? (1913)

Yes. In 1913, Georges Sagnac (1869-1928), a French physicist, performed an experiment to test whether light was a wave or a particle.[151] Like Maxwell,[152] and many other theoretical physicists, Sagnac thought the wave nature of light required a medium necessary for light transmission. Like the media for sound, this medium controlled the velocity of wave motion through it. The constant velocity of sound through air is 343 m/s while the constant velocity of light through "vacuum" is 300 million m/s. In addition, this medium, variously called "aether" or "ether," consisted of particles so tiny[153] they permeated everything.

Proof that light is a wave

Because light velocity is so great, Sagnac used an interferometer, which can show an interference pattern or "fringe" on photographic paper when it detects that light from two different directions is not in phase (Figure 30). This fringe is sort of like the fuzziness you see when your camera is out of focus.

---

[150] Borchardt, 2011, Einstein's most important philosophical error.
[151] Sagnac, 1913, The demonstration of the luminiferous aether by an interferometer in uniform rotation; Sagnac, 1913, On the proof of the reality of the luminiferous aether by the experiment with a rotating interferometer.
[152] Shaw, 2014, Maxwell's aether.
[153] Possibly with diameters as small as 10^{-19} cm and a mass of 10^{-47} g according to my speculations calculated from Planck's constant, which concerns the "smallest unit of motion." See the details in the section on Aether Deceleration Theory.

194   *Infinite Universe Theory*

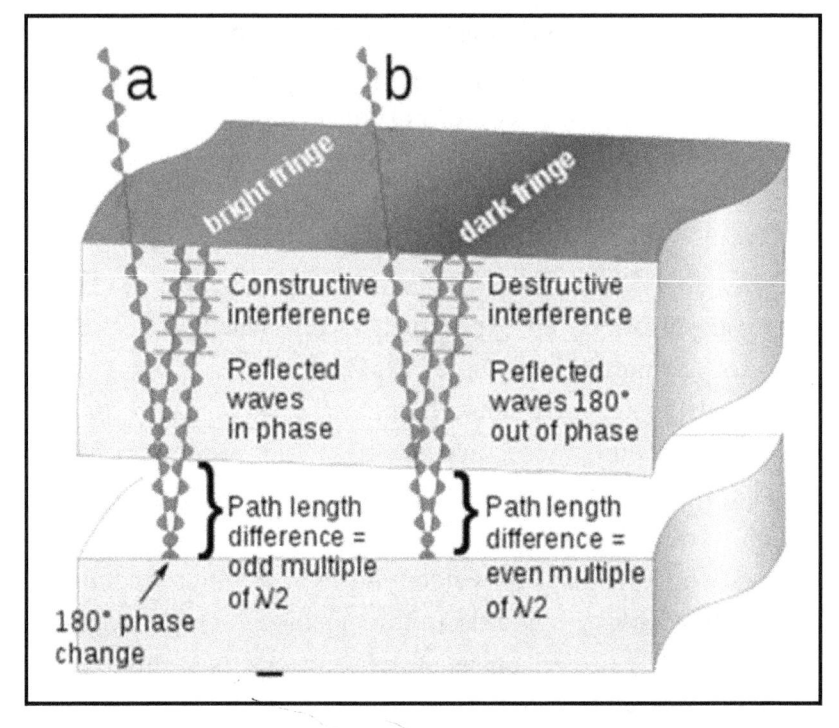

Figure 30 "How interference works. The gap between the surfaces and the wavelength of the light waves are greatly exaggerated. The distance between the dark fringe (a) and the bright fringe (b) indicates a change in the gap's thickness of 1/2 the wavelength."[154] Credit: Chetvorno.[155] Note that the path length on the bottom left is a half cycle longer than the path length on the bottom right. If the two lengths were identical, there would be no fringe.

    The setup is a bit complicated. *Wikipedia* gives the details if you need them.[156] Figure 31 shows the gist of the experiment. It involved a rotatable apparatus around which Sagnac sent light in opposite directions within a self-contained system. He attached both the light source and the detector to the apparatus, so the distance between them would be the same whether or not the

---

[154] http://go.glennborchardt.com/fringes.
[155] (Own work) [CC0], via Wikimedia Commons. Accessed on 20170221.

[156] http://go.glennborchardt.com/sagnac

platform was rotating (Figure 31, left side). Upon rotation, light traveling in the direction of rotation takes longer to reach the detector, because the detector is always moving away from it (Figure 31, right side). Light traveling opposite to the direction of rotation arrives at the detector sooner because the detector is moving toward it. The experiment has been refined and repeated frequently, always with the same outcome: a fringe (Figure 30) always shows up.

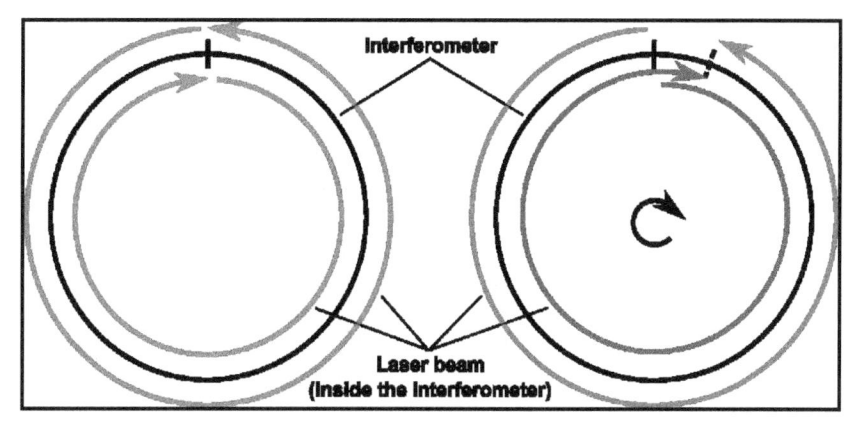

Figure 31 "When the interferometer starts to rotate clockwise the clockwise propagating laser beam has to cover more distance because the detector is moving away. The counter-clockwise propagating beam has to cover less distance as the detector is moving towards it. Now the two laser beams have a different phase at the point where they interfere resulting in a different amplitude of the signal at the detector."[157]
Credit: Berhard Albrecht.

Since there is much ingrained resistance to considering aether as a medium, let us consider a different medium, water. Water waves are analogous to light waves in that their velocity, like the velocity of light, is dependent solely on the properties of the medium. Suppose we have an Olympic-size swimming pool that

---

[157] Albrecht, 2018, Sagnac interferometer.

is 100 m long (Figure 32). Next, suppose we can move the right end of the pool back and forth. Thus, if the velocity of a water wave is 2 m/s, then the wave will travel 100 meters to the end of the pool in 50 seconds. If we start the wave and then move the right end of the pool (the target) to be 110 meters away, then it will take 55 seconds for the wave to hit the end of the pool. If we move the target to be only 90 meters away, then it will take 45 seconds. That is what Sagnac did in his experiment with light. Once light left the source, it traveled a longer distance when the detector moved away from the source and a shorter distance when the detector moved toward the source (Figure 31). The difference showed up as an interference pattern or fringe. In both cases, the medium acts on its own independently of its surroundings, with wave motion traveling through it at a constant velocity. Pretty simple.

Proof that light is not a particle

If light was a particle, the result would be completely different—there would be no fringe. That is because the velocity of a particle includes the velocity of its source. Thus, if one threw a ball from the front of the pool to the end of the pool at a velocity of 100 m/s, it would arrive at the end of the pool in one second. Now, if the pool was on a cruise ship traveling at 10 m/s, both the front of the pool and the end of the pool would be moving at 10 m/s. That means the ball would be moving at 10 m/s immediately before the throw, giving it a velocity of 110 m/s as a result of the throw. In one second, the ball would travel 110 m, but the end of the pool would move away by 10 m during that one second. The ball would travel 110 m, precisely hitting the end of pool after that one second. If the ship was traveling in the opposite direction, both the front and end of the pool would move in the opposite direction. In that case, the velocity of the front of the pool would be subtracted, yielding a ball velocity of 90 m/s. In one second the end of the pool would move toward the source of the throw by 10 m. Again, the ball would hit the end of the pool precisely. If the ball was the hypothesized light particle, no interference pattern or fringe would be seen.

As seen by the words "Proof of the Reality of the Luminiferous Aether" in his title, it is obvious Sagnac thought of his experiment as a straightforward proof of the existence of aether.[158] And indeed it is. Nonetheless, its interpretation is not at all clear to indeterminists, who wasted time debating the Sagnac experiment for over a century. The indeterministic regression had been set in motion with some particularly wild claims made by Einstein.

Ad hoc hypotheses and special pleading

From Sagnac's experiment it was obvious light was not a particle, but a wave in a sea of aether particles, just as sound was a wave in a sea of nitrogen and oxygen particles. Stemming most likely from his 1905 Nobel Prize work on the photoelectric effect,[159] Einstein became convinced light was a special particle existing as a sort of unprecedented "wave-packet." He probably breathed a sigh of relief when Sagnac showed light was not a particle in the classical sense. He once wrote of the embarrassment he would face if anyone could show light was a classical particle:

> *If the speed of light depends even in the very least on the speed of the light source then my whole theory of relativity, including my gravitational theory is false.*[160]

---

[158] This shows the acute importance of titles. Sagnac's title portrays the truth ("proof of...ether"); Hubble's title did not ("relation between distance and radial velocity").

[159] http://go.glennborchardt.com/PE; Einstein, 1905, Concerning an Heuristic Point of View Toward the Emission and Transformation of Light.

[160] Einstein, 1913, Letter from Einstein to Freundlich.

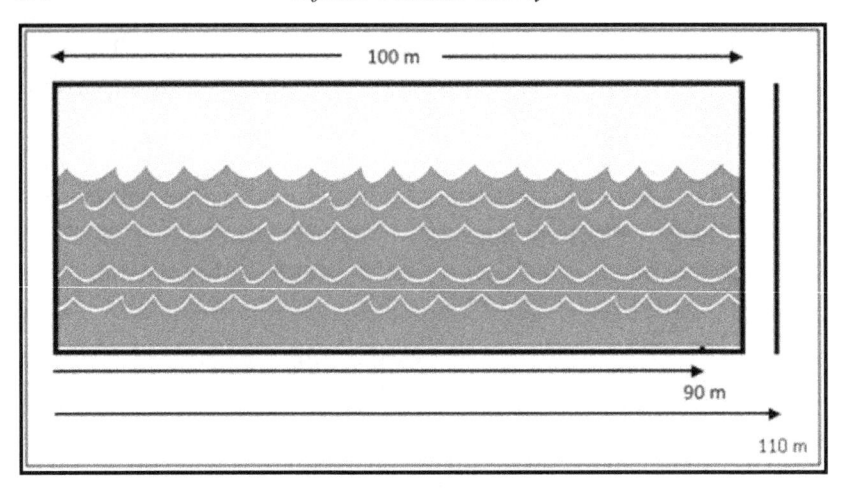

Figure 32 Olympic pool with movable end.

Of course, this did not remove the part of the Sagnac experiment proving light was a wave instead. One could dispute Sagnac's classical interpretation that all waves must occur in a medium only by resorting to some "special pleading" as Einstein did in developing his Special Relativity Theory. Special pleading is an attempt to get around falsifying results by asserting the phenomena involved are special, and therefore not subject to previous generalities. In science, the introduction of ad hoc additions to theory usually is frowned upon (Figure 33). Einstein's special pleas were dramatic, so-called "revolutionary" rejections of mechanics and its philosophical foundation. As mentioned, light speed was constant because light was a wave in the aether. To treat light as a particle in the face of that, he had to make some special pleas (Table 6).

Not having a proper BS meter, indeterminists accepted all of Einstein's ad hocs, boosting the regressive movement in physics. Recognizing the absurdities, reformists tended to accept some of the ad hocs and not others. For once, we agree with Einstein; light is not a particle in the classical sense. Nonetheless, Einstein's special pleading made no sense to classical mechanists. One primary objector was Walter Ritz, a young

Swiss theoretical physicist, who, in 1908, tried to reclaim Newton's corpuscular theory of light.[161] This reiteration became known as the "Ritz theory," emission theory,[162] emitter theory, or ballistic theory. In each case, theorists considered light as a classical particle, which supposedly took on the motion of the source (v) in the classical way (e.g., c+v). This was before Sagnac finally proved it did not. That led to much doubt in 1913, and emission theory has been thoroughly discredited since,[163] with the death knell given by Brecher in 1977.[164] The fact the theory was still taken seriously for 64 years after its demise shows how difficult it is to rid physics of falsified theories. Even John Chappell, one of the organizers of the dissident physics movement in the US, once wrote:

> ...I do not think a complete understanding of Sagnac's results is possible without realizing that the Ritz approach to electromagnetic theory is essentially correct and represents the main path for the future development of physics.[165]

This is an excellent example of the diffidence and caution maintained by those facing an overwhelmingly popular paradigm. Dissidents commonly try to reform physics by objecting only to certain aspects of relativity while trying to avoid the "A" word altogether.[166] In this case, aether enthusiast Chappell was supporting both the particle theory and the wave theory, sort of like the regressive agnostics who still support wave-particle dualism.

---

[161] Ritz, 1908, Recherches critiques sur l'Électrodynamique Générale.
[162] http://go.glennborchardt.com/ET
[163] Fox, 1965, Evidence Against Emission Theories; Martínez, 2004, Ritz, Einstein, and the Emission Hypothesis.
[164] Brecher, 1977, Is the Speed of Light Independent of the Velocity of the Source?
[165] Chappell, 1965, Georges Sagnac and the Discovery of the Ether.
[166] Dowdye, 2012, Discourses & Mathematical Illustrations Pertaining to the Extinction Shift Principle.

200   *Infinite Universe Theory*

Nevertheless, Sagnac's result *was* consistent with Einstein's ad hoc postulate that if light is a particle, it does not take on the motion of its source. Einstein also proclaimed the speed of light would be the same for all observers regardless of their motion. There has been much resistance and confusion over this point. Because light is a wave, its velocity is dependent on the medium. Neither the velocity of the source nor of the observer is relevant. Light will always travel at *c*. However, attempts to measure light speed must take into account the *position* of the observer relative to the source as in the Doppler Effect. We can see this from the fact that Sagnac, Einstein, and most everyone else agree the velocity of light is more or less constant. If we did not get that result, then there would be something wrong with our calculations. Wave motion through any medium is always relatively constant as long as the properties of the medium remain relatively unchanged. The idea that the velocity of light usually is constant therefore is no great revelation.

Figure 33 Sid Harris's classic cartoon illustrating scientific skepticism toward an ad hoc. Credit: ©Sidney Harris (sciencecartoonsplus.com).[167]

---

[167] http://sciencecartoonsplus.com/index.php

As in the aether explanation, moving the target closer to the source will produce a short path and moving the target away from the source will produce a long path. As in the water analogy, the time for the long path would be 55 seconds and the time for the short path would be 45 seconds. Sagnac got fringes because light from each path showed up at the detector at slightly different times, as indicated by the overlap in the long gray line in Figure 31. After all, that is what it means for waves of constant length to be out of phase. The fact the targets or "observers" were in motion was irrelevant once we calculated the positions of the targets. Here is another illustration from baseball:

Suppose a pitcher is throwing a 100-mph fastball at the catcher. If the catcher runs toward the pitcher, the velocity of the fastball is unaffected. Running at 10 mph, the catcher would be about 54 feet from the pitcher when the ball arrived. A pitch normally would take about 0.41 seconds to travel the 60 feet to the plate, but at a distance of only 54 feet, it would take less than 0.37 seconds. The critical measurements involve only the distance traveled and the time it took. Whether the catcher was running or standing still at 54 feet, the result is the same. The measurements of the velocity of light and the velocity of a fastball, are unaffected by the motion of the observers.

The upshot is that the Sagnac effect provides support for aether and falsifies the theory that light is a particle. The only way to save the particle theory was to give it miraculous properties. Indeterminists who accepted Einstein's eight ad hocs mentioned above could view the Sagnac effect, instead, as support for relativity. The mistaken belief that Sagnac supports both theories has provided endless opportunities for reformists and regressives to debate the issue. Ad hocs tend to do that. Folks tend to choose sides having nothing to do with the evidence. If we used *Occam's razor*,[168] the aether theory would win easily. Its simplicity and conformity with classical mechanics makes it by far the most elegant solution. Theories should not be more complicated than necessary, for each addition requires substantiation. The upshot is that Einstein's light particle has all sorts of unprecedented supernatural properties: it is massless, exhibits instantaneous acceleration, always travels at a constant speed, never encountering another light particle, and has both particle and wave properties at the same time. Einstein's special pleading was not a problem for regressive physicists. Is it a problem for you?

15.2 Did de Sitter prove aether exists? (1913)

Yes. 1913 was a "nonmiracle" year for falsifying relativity. The reaction to Einstein's outrageous ad hocs (Table 6) not only included Sagnac's experiment, but it also included de Sitter's famous observations from astronomy.[169] His work was an uncomplicated reaction (1 page) to Ritz's complicated claim that light was a classical particle (130 pages). The observations were quite simple. Some star systems exist in pairs called "binaries." Being similar in mass, each star revolves around the other. That means one star is always coming toward Earth (A) and one is always going away (B) (Figure 34). If light was a classical particle, it would take on the motion of its source if it was coming toward us (c+u) and if it was going away from us its motion would be less (c-u). Light from A would arrive before light from B. That did not happen. According to de Sitter, "Ritz's accepted dependence of the speed of light on the movement of the source is absolutely inadmissible."

---

[168] https://en.wikipedia.org/wiki/Occam's_razor
[169] de Sitter, 1913, An Astronomical Proof for the Constancy of the Speed of Light.

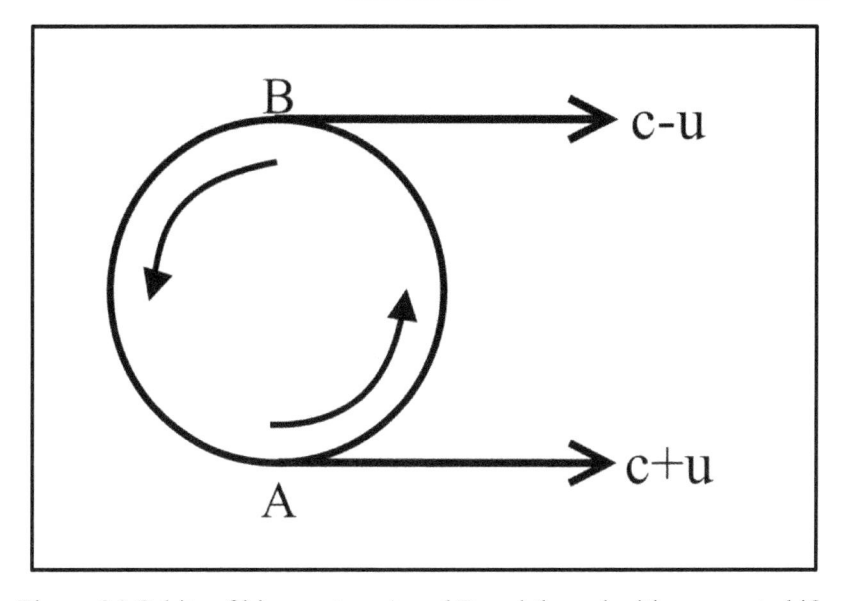

Figure 34 Orbits of binary stars A and B and the velocities expected if light was a particle.[170]

All this means is that light is a wave traveling through a medium (aether), that, like all media, conducts motion at a constant velocity. Like Sagnac's experiment, this put the kibosh on the theory that light was a particle. Again, the only way to revive the particle theory was to claim light was a special particle that adhered to Einstein's eight ad hocs (Table 6). Once again, physicists had a choice: accept deterministic aether or accept indeterministic ad hocs. Note that indeterminism was already taking hold at that time even though Einstein's anointment would not take place for another 6 years. Unlike Sagnac, whose title straightforwardly claimed he had found "Proof of the Reality of the Luminiferous Aether," de Sitter was careful to make no mention of aether. Instead, he decided to go with "Proof for the Constancy of the Speed of Light," which spoke to the truth as much as it demurred on the issue at hand. De Sitter would take no position on the determinism-indeterminism philosophical

---

[170] Ibid. Figure from de Sitter.

struggle, which, in this case was masquerading as the aether vs. ad hocs controversy.[171]

Although the works of Sagnac and de Sitter present an open and shut case for aether, reformists continue their debates about whether the velocity of light is dependent on the velocity of its source. There are arguments about reference frames, when the only important reference frame is the aether medium itself. They have a clear choice many are reluctant to make: aether or Einstein's adhocs.

15.3 Did Eddington prove space was curved? (1919)

No. It was not until Eddington's famous eclipse observations that Einstein was anointed the world's foremost genius.[172] Frankly, like relativity itself, the hypothesis tested by those observations was not exactly clear. As mentioned, Special Relativity Theory claimed light was a corpuscle[173] or particle that, like all particles, should be affected by gravitation despite its being massless. On the other hand, General Relativity Theory claimed the space[174] surrounding massive bodies was curved despite being assumed perfectly empty. During an eclipse, it becomes possible to view what happens to light from far away stars when it passes the Sun. Although the instruments he used were not up to the task, Eddington nonetheless reported that passing light indeed was bent toward the sun.[175]

---

[171] As part of the struggle, the choice was not as clear as I have painted it. As far as I can tell, there never was an "aether vs. ad hocs controversy" per se. I have never seen anything like Table 6.

[172] Dyson, Eddington, and Davidson, 1920, A Determination of the Deflection of Light by the Sun's Gravitational Field, from Observations Made at the Total Eclipse of May 29, 1919.

[173] Named a "photon" by Lewis in 1926 (The conservation of photons).

[174] Sometimes construed as "space-time," although the experiment necessarily could only test space.

[175] Dyson, Eddington, and Davidson, ibid.

This supposedly astounding result made headlines all over the world. The 42-page paper published by the Royal Society was filled with data. It is worth it to get into some of the details, which I will simplify as much as possible. General Relativity Theory predicts light bending should be an inverse function of distance from the Sun (1.75 arcsec/R, where R=radius of the Sun). There were always suspicions about Eddington's work, but in 2009, Dr. Edward Dowdye, a retired NASA physicist, showed nothing of the sort occurs (Figure 35).

According to Dowdye, Einstein's predicted angular deflection of 1.75 arcsec is observed with modern instruments only at the solar plasma rim. The plasma rim is a thin veneer less than 2.5% of the diameter of the Sun, which would be <0.8 arc minute. Eddington could not measure that directly because of the Sun's corona (Figure 36). Thus, at a distance of 3.17 radii from the center of the Sun (50 arc minutes), General Relativity Theory predicts the deflection should be 0.55 arcsec (1.75 arcsec/3.17). Eddington reported 0.625 arcsec.[176] As claimed by Dowdye, modern instruments find no deflection at that distance. It is not clear how Eddington was able to get that result, along with all the other deflections he reported. There is speculation his erratic data allowed him to get it by "cherry picking," that is, by selecting only data that agreed with what he wanted to prove. That might be de rigueur in theology, but it is supposed to be a cardinal sin in the scientific community. Nonetheless, indeterminists of the day tended to overlook this. Others subsequently even called it "the greatest hoax in 20th century science."[177] In any case, we should never use the Eddington paper in support of relativity, as many still do today. I repeat: Recent work has failed to confirm the claims, finding instead,

---

[176] Ibid, pp. 306 and 322.
[177] Moody, 2009, The eclipse data from 1919; Almassi, 2009, Trust in expert testimony; Falkenstein, 2010, Eddington's Experiment Was Bogus.

there is absolutely no light bending beyond the plasma rim. The upshot is that relativity has been falsified once again. Light is not a particle that responds to either gravitation or the imagined curved empty space. Instead, light is a wave that has a curved path during refraction, which occurs whenever it enters an atmosphere. The Sun obviously has an atmosphere (Figure 36).

The Eddington observations were the first of many misinterpreted experimental "proofs" of relativity. I put the word "proof" in quotes for a special reason. Indeterminists were so enthusiastic about the prospects for finally defeating mechanism that many of their so-called proofs of relativity relied on either suspect data or interpretations of the data from the indeterministic viewpoint. Einstein was especially lucky at making predictions, many of which came true, but for the wrong reasons.[178] In this instance, the 1.75-arcsec result was due to refraction in the Sun's immediate atmosphere. Of course, the Sun's atmosphere, like Earth's atmosphere, contains baryonic matter under gravitational influence. The density (e.g., pressure) of a planet's atmosphere is dependent on its rotational period, radius, and mass.[179] Einstein's equation on Figure 35 only considered radius and mass. Anyone still not convinced Eddington's results were bogus could show the refraction was a function of rotational period as well.

---

[178] Waller, 2002, Einstein's luck. [See also Einsteinism.]
[179] Puetz and Borchardt, Universal Cycle Theory, ibid, p. 138.

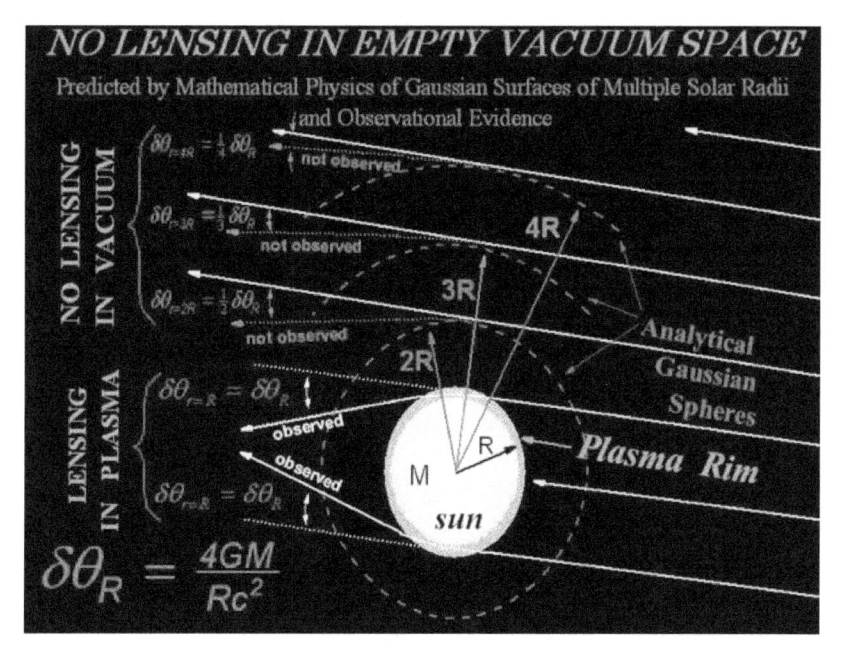

Figure 35 Light waves from distant stars bend only in the plasma rim of the Sun due to refraction. They are unaffected by gravitation, contrary to the predictions of relativity (from Dowdye, 2012).[180]

So there it is. Einstein's treatment of light as a corpuscle in Special Relativity Theory and of time as an object in General Relativity Theory led to an observation thought to confirm both theories though it did no such thing. The indeterministic idealism and faulty interpretations used in both theories remain essential parts of regressive physics a full century later. Amazingly, simple atmospheric refraction was used as celebratory support for indeterministic interpretations of theories that, in turn, provided the foundation of the Big Bang Theory. None of this would have happened if classical mechanics had dropped its obsession with *finity*, evolving directly into neomechanics. That was not possible for sociological reasons to become clearer later.

---

[180] Dowdye, 2012, Discourses & Mathematical Illustrations Pertaining to the Extinction Shift Principle. [See also Dowdye, 2010].

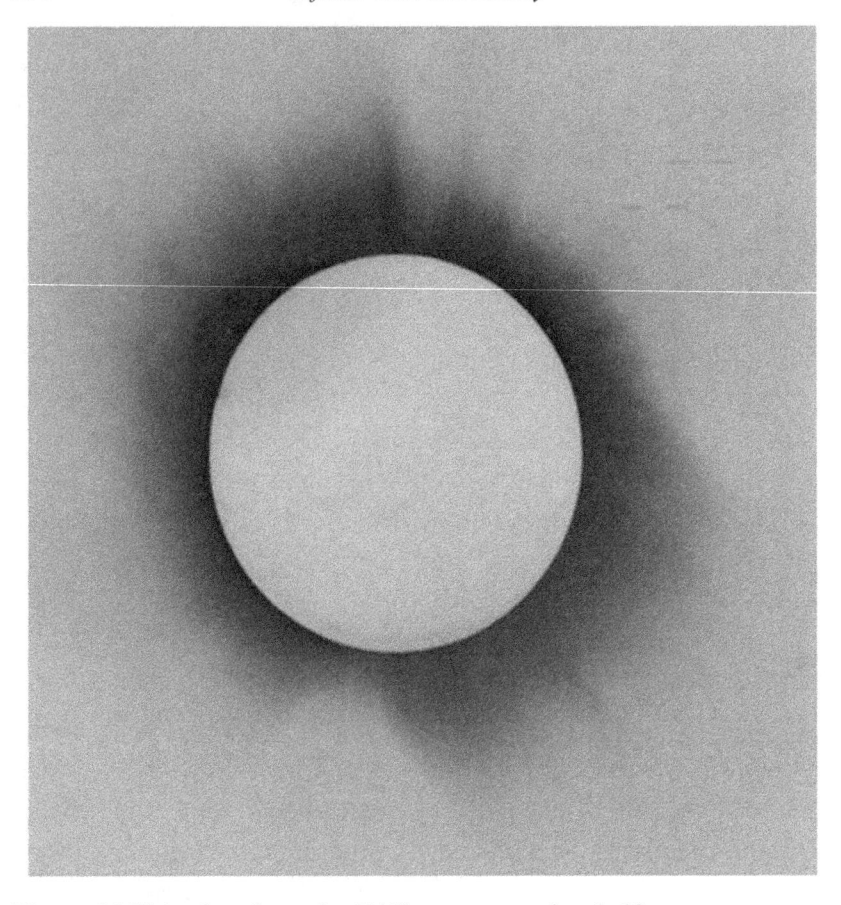

Figure 36 This plate from the Eddington paper is a half-tone reproduction from one of the negatives taken with a 4" lens at Sobral, Brazil. The corona prevented any observation of light bending in the plasma rim at the surface of the Sun (from Dyson, Eddington, and Davidson, 1920).[181]

15.4 Did Eddington prove light is affected by gravitation? (1919)

No. As seen above, the bending of light Eddington thought he observed was due to simple refraction in the Sun's atmosphere. Not only was that phenomenon not due to Einstein's

---

[181] Dyson, Eddington, and Davidson, ibid, Plate 1.

"curved space-time," but it also was not due to gravitation.[182] As mentioned, to get around all the evidence light was wave motion in the aether, Einstein had to assume his imagined light particles were massless. But all real microcosms contain submicrocosms giving them mass, which is affected by gravitation. Logically, if you still believe that Eddington proved light was affected by gravitation, then you cannot also believe Einstein's ad hoc that light was massless.

15.5 Does the gravitational redshift confirm relativity? (1960)

No. What has been misnamed the "gravitational redshift" occurs when light leaves a massive cosmic body; conversely, a "gravitational blueshift" occurs when light approaches a cosmic body. This is one of the most important and most popular Einsteinisms claiming to confirm relativity. By using his corpuscular theory of light and the attraction theory of gravitation, Einstein predicted photons would gain energy under the "pull" of gravity, becoming blueshifted. Photons leaving a gravitational field would lose energy as they "fought against gravity," becoming redshifted. In 1960, a famous experiment performed by Pound and Rebka in a 22.5-m tower at Harvard showed electromagnetic radiation indeed was blueshifted when directed down the tower and redshifted when directed up the tower.[183] Many subsequent experiments confirmed the effect, with electromagnetic radiation emitted from all sources, such as galaxies and even galactic clusters being redshifted as it left the vicinity of those massive objects.

---

[182] Dowdye, 2010, Findings convincingly show no direct interaction between gravitation and electromagnetism in empty vacuum space.
[183] Pound and Rebka, 1960, Apparent Weight of Photons.

Refraction

There are many reasons for redshifts and blueshifts. One of them is due to variations in the elasticity and density of the medium per the Newton-Laplace equation. An elastic medium generally conducts wave motion faster than a less elastic medium; a dense medium generally conducts motion slower than a less dense medium. The equation uses the ratio between elasticity and density. Sound travels at about 340 m/s in air, 1,500 m/s in water, and 5,000 m/s in iron, not because of increasing density, but because of increasing elasticity, which overwhelms the density factor. Elasticity is a measure of the degree of interconnection between the atoms or molecules, with gases generally being less elastic than liquids, and liquids being less elastic than solids. Seismic wave velocity also is a function of elasticity, which increases with depth in the Earth due to overburden pressure.[184] The main reason for this is the closeness of the particles involved. That is why you can hear a train coming 15 times sooner by putting your ear to the railroad track. As velocities increase, wavelengths get longer—the waves are "redshifted." That is why your voice has a much lower pitch in water than in air. NOAA puts it this way:

> ...the wavelength of a sound equals the speed of sound in either air or water divided by the frequency of the wave. Therefore, a 20 Hz sound wave is 75 m long in the water (1500/20 = 75) whereas a 20 Hz sound wave in air is only 17 m long (340/20 = 17) in air.[185]

As mentioned, light travels at nearly 300,000,000 m/s in air, but at only 225,000,000 m/s in water. This happens despite water's higher elasticity (increased velocity) and greater density (slower velocity). If we use the sound analogy, the Newton-

---

[184] http://go.glennborchardt.com/seismic
[185] http://go.glennborchardt.com/sound

Laplace equation would predict that light should travel almost five times as fast in water as in air. That does not happen because, according to Infinite Universe Theory, the medium for light is aether, not water. The elasticity of aether probably is extremely low (interconnection would be low, as it is in rarefied gases) and likely does not change much throughout the universe. This means that the Newton-Laplace equation for aether would be dominated by density. Per Aether Deceleration Theory (Chapter 16.3) all baryonic matter is surrounded by decelerated aether particles. Gravitation leaves behind decelerated aether particles that increase the density of the aether medium at the same time as its pressure or activity is reduced. Similar to the NOAA example above, the wavelength (λ) of light traveling from air to water is reduced according to the equation:

$\lambda_{water} = \lambda_{air}/n$ [1]

Where:

λ_{water} = wavelength in water

λ_{air} = wavelength in air

n = refractive index

The *refractive index* (n) for water is (300,000,000 m/s)/(225,000,000 m/s)=1.333. The wavelength of light is shortened to 75% of what it is in air. Thus, red light with a wavelength of 650 nm[186] in air has a wavelength of 488 nm in water. As in the NOAA example, the number of cycles per second (frequency), however, remains unchanged. Because the light is slower, it travels less distance during each cycle, resulting in the shorter wavelength. As an example, if a 100 cycle/s wave traveled in air at about 300,000,000 m/s, it would be 3,000 km

---

[186] nm =nanometer, 0.000000001 meter (a billionth of a meter or 10^{-9} m)

long. If the 100 cycle/s wave traveled in water at 225,000,000 m/s, it would be 2,250 km long. After leaving the water for the aether-rich air, the wave once more would be 3,000 km long. Again, during the transition from water-to-air, the number of cycles/s (frequency) does not change—only the velocity of light would change due to the decreased aether density in air. Now, this point about frequency will become important when I discuss the real reason for Pound and Rebka's result (it was not due to refraction, but you need to know about refraction anyway). Note that the color of light depends on the frequency, not the wavelength—the 488 nm red light in water is just as red as the 650 nm red light in air. Thus, red laser light entering and exiting a glass of water remains red throughout the process.[187] Note again none of these changes in wavelength has anything to do with color. The "red" in "redshift" merely is our shorthand way of indicating the wavelength has become longer. It stems from the fact red light in air has a longer wavelength (650 nm) than blue light in air (475 nm). We might even use that terminology to describe the Doppler Effect, with the sound of the train whistle appearing to become "redshifted" after it passes by.

Now it is obvious a small amount of water in the atmosphere would produce a decrease in wavelength—a "blueshift," if you will. Any lessening of water content would produce an increase in wavelength—a "redshift." Light coming from any cosmic body with an atmosphere would have a "gravitational redshift" when it speeds up as it traverses its thinning atmosphere. At best, however, this would be only an indirect effect of gravitation. It would not be the Einsteinian "fight against gravity." At most, one might consider it a fight against the molecules in the atmosphere held there by gravity. As in all refraction, light is

---

[187] You can prove this yourself by putting a pencil in a glass of water and shining your red laser pointer through the glass and onto the pencil. The dot on the pencil will be red.

absorbed and emitted whenever it collides with baryonic matter. Above all, none of this proves light is a particle affected by gravitation. Again, these effects are merely the result of refraction.

To be clear, let me repeat. What *is* affected by gravity is the diluting agent contaminating the aether medium through which the light wave travels. As in the water example, the aether in Earth's atmosphere is made denser and its bulk modulus decreased by the baryonic matter—mostly nitrogen, water and oxygen. Because there is less aether pressure and more aether density, light will travel slower near sea level than at high altitude where there is greater aether pressure and less aether density. In addition, as I speculated in Chapter 16.3 on gravitation, aether pressure/activity tends to decrease with nearness to massive bodies due to aether particle deceleration. Thus, if light speed is a function of aether pressure, then light leaving a massive body will travel faster than light arriving at a massive body. This is exactly what Pound and Rebka discovered. Waves going away were slightly longer than waves coming toward Earth. In this, Einstein once again got lucky. Even though light is not a particle, as he assumed, gravity increases baryonic concentrations in the air and thereby densifies the aether and decreases its bulk modulus necessary for the transmission of light. Aether particles collide with this baryonic matter, becoming decelerated and concentrated. This appears as an increase in aether density, which according to the Newton-Laplace equation must result in a slowing of light speed. Thus, for instance, any humidity in the atmosphere will cause light to slow down just as it does in water.

Aether Pressure Gradient

Now, any "gravitational redshift" produced by refraction due to Earth's atmosphere would be tiny. The index of refraction of the atmosphere at sea level is only 1.000277. This means light under natural conditions in the atmosphere travels at (300,000,000 m/s)/1.000277=299,917,000 m/s—0.0277% slower than in vacuum. For yellow light with a wavelength of 589 nm, the reduction in velocity due to Earth's entire atmosphere would produce a reduction in wavelength of only 0.163 nm, which normally would be undetectable. Nonetheless, the Mossbauer setup used by Pound and Rebka was so sensitive they were able to detect what they thought to be a *frequency* change of only 1 part in 10^{15} for the gamma rays they used across the 22.5 m. Although frequency never changes, the wavelength did, as it does in pure water. This would be a reduction in the 589-nm wavelength of sodium light by only 5.89 X 10^{-13} nm. The wavelength of light going away from Earth would be increased by a similarly tiny amount, as the velocity of light increases due to increasing aether pressure and decreasing density. Now, Pound and Rebka were exceedingly clever in that they used flowing helium gas to displace the atmospheric gases that would have produced refraction, masking effects they thought due to gravitation. Thus, their measurement involves no significant refraction because the refractive index of helium gas is negligible. In any case, helium would not have a significant gravitational gradient within the 22.5 m used in the experiment. But, according to our Aether Deceleration Theory, aether pressure/activity always increases with distance from baryonic matter no matter how slight the distance. There is always a Gravitational Pressure Gradient surrounding all objects. That is, after all, what produces gravitation. This means the gravitational redshift is produced by two factors: 1) natural refraction in the atmosphere, 2) increases in distal aether pressure and increases in proximal aether density due to aether particle deceleration.

The result is a redshift indeed, but it is not a direct effect of gravitation as Einstein supposed. That always was a big contradiction, because, according to Special Relativity Theory, his imagined light particle was assumed massless. Of course, as mentioned, a massless particle could never be affected by gravitation—it would not even obey Newton's equation for gravity ($F = Gm_1m_2/r^2$). Obviously, if m_1 was zero, then F would be zero.

Again, the gravitational redshift is simply due to velocity increases as light enters areas of increasing aether pressure far from baryonic matter. That same petard bedeviled Eddington's 1919 "proof" empty space was curved and light was affected by gravity. As in the Eddington experiment, it certainly is no proof light is a particle, as erroneously interpreted by Pound and Rebka and Einstein's other followers. The equation for the experiment can be expressed as:

$c/f = \lambda$ [2]

Where:

c = velocity of light, 300,000,000 m/s

f = frequency, cycles/s

λ = wavelength, nm

This implies that, if c is assumed constant, then frequency must increase whenever wavelength decreases. If c should decrease due to refraction, as it does in water, then the wavelength would decrease by the corresponding fraction while the frequency[188] would not. Now, the Mossbauer setup does not distinguish between changes in wavelength and frequency. Of course, Pound and Rebka knew frequency does not change during refraction and ruled out wavelength changes due to refraction by using the helium. Being in complete aether denial,

---

[188] http://go.glennborchardt.com/frequency

however, the options for explaining the wavelength shifts they observed were thus limited to relativity. As you probably know, relativity assumes the velocity of light in a "vacuum" (or in helium for that matter) is constant. With light velocity and wavelength assumed constant, the changes observed in the experiment had to be ascribed to *frequency*.

Aside from the Doppler Effect, in classical mechanics and in neomechanics there is no known physical reason for frequency to change once electromagnetic radiation is set into motion. That was confirmed by the experiments mentioned above clearly showing the frequency of light traveling from air to water does not change—only the wavelength and velocity changes. Frequency is cycles divided by seconds. As a relativist and aether denier, the only way you could explain your assumed frequency changes would be to consider time to be a variable. That is, if you assumed time was matter and could dilate, then you might also assume each second would take slightly longer than normal. Dividing the number of cycles by a bigger number would result in a decrease in frequency. Now look at equation 2 again. Dividing c by a lower frequency would result in an increase in wavelength: the gravitational redshift. The reported gravitational blueshift would be a result of "time contraction." All this is consistent with Special Relativity Theory and General Relativity Theory, and is habitually cited as a rock-solid confirmation of relativity, but it is completely wrong.

The neomechanical explanation is rather simple. Look at equation 2 once again. Suppose c was not constant. Then, a slight increase in c would result in a slight increase in wavelength. Light traveling faster away from a massive body would be "redshifted." Light traveling slower toward a massive body would be "blueshifted." That is exactly what the Pound-Rebka data showed. Note that even these terms are misnomers because, without the frequency change assumed by the relativists, there really is no change in color. As mentioned before, the much-ballyhooed "redshifts" of cosmogony are the result of changes in wavelength, not of frequency. That is why many galaxies near the edge of the observable universe are not red despite their high so-called "redshift" values (Figure 9). They are called redshifts merely because the waves involved have lengthened toward the red (long wavelength) end of the electromagnetic spectrum (Figure 11). As mentioned, color is determined by frequency, not wavelength.

The upshot is if one erroneously assumes time can dilate, each second would be slightly longer than normal, the frequency would then remain the same and so would the velocity of light. Even though both time dilation and length contraction originated with Lorentz, Einstein generally gets the major share of the blame.[189] Both dilation and contraction are invariably used to protect the constancy of c and maintain the ruse that light does not need a medium. The upshot on Pound and Rebka: good data; bad interpretation.

15.6 Did clocks flown around Earth confirm relativity? (1972)

No. The famed Hafele-Keating (H-K) experiment alluded to in most every physics class confirmed that equations used in the Special and General Relativity Theories were valid.[190] Relativistic time differences, like all motion differences in the Infinite Universe are now a fact. Nonetheless, reformists have been reluctant to accept any experiment as confirmation of relativity. Their complaints take many forms, but the H-K experiment has been particularly vulnerable. That is partly because Hafele and Keating did not publish the raw data and the

---

[189] Schlafly, 2011, How Einstein ruined physics.
[190] Hafele and Keating, 1972a, 1972b, Around-the-World Atomic Clocks: Predicted and Observed Relativistic Time Gains.

results were not straightforward. For instance, clocks traveling east around Earth lost time and those traveling west gained time. A simple-minded test of the Lorentz factor γ[191] through perfectly empty space would have shown "time dilation" to be dependent only on the velocity of the plane, which would have been the same in both directions.

Reformists have attacked the H-K data and the manipulation they performed in confirming relativity.[192] Those details were unknown until 1996 when Spencer and Shama finally were able to get the raw data from Keating. Kelly did an especially detailed analysis of those results, pointing out their highly erratic nature—some clocks lost time, while others gained time even though they were sitting side-by-side on the same plane (Table 7). That slow westbound clock lost 44 ns, while a clock right next to it on the plane gained 413 ns—hardly a spread commensurate with the 10% error claimed in the 1972 Science article. It also seemed impossible that any appropriate fudge factor could be applied to both measurements to get them to agree. But there are standard procedures to do just that, as Hafele and Keating explained on page 169 of their Science paper. The gist of that is the observation that Cesium clocks do not simply keep time smoothly—they take wholesale jumps at irregular intervals. But, according to Kelly, there was no reason other than instrumental error for the raw data to appear as messed up as they were. Here is his conclusion:

> *The H & K tests prove nothing. The accuracy of the clocks would need to be two orders of magnitude better to give confidence in the results. The actual test results, which were not published, were changed by H & K to give the*

---

[191] $1/\sqrt{1 - (\frac{v^2}{c^2})}$.

[192] Spencer and Shama, 1996, A new interpretation of the Hafele-Keating experiment; Kelly, 2000, Hafele & Keating Tests: Did They Prove Anything?

impression that they confirm the theory. Only one clock (447) had a fairly steady performance over the whole test period; taking its results gives no difference for the Eastward and the Westward tests.[193]

Now, Kelly's analysis has been considered incorrect, naïve, and certainly not sophisticated enough for the experimental methodology. According to physicist T.J. Roberts at Fermilab:

His criticism does not stand up, as he does not understand the properties of the atomic clocks and the way the four clocks were reduced to a single "paper" clock. The simple averages he advocates are not nearly as accurate as the paper clock used in the final paper—that was the whole point of flying four clocks (they call this "correlated rate change"; this technique is used by all standards organizations today to minimize the deficiencies of atomic clocks).[194]

I have to admit that I was taken in by Kelly's critique, even using it as an example in my paper on Einstein's objectification of motion.[195] At the time, I was ready for any explanation of why time cannot dilate because that did not make any sense to me. I was not even interested in the more recent experiments showing the time differences to be real.[196] I should have been more skeptical of the skeptics.

Fact is, atomic clocks have gotten ever more accurate and precise. By using the "gravitational redshift," they now can be used to measure altitude within 10 to 1 m. With 10^{-18} accuracy in the developmental stage, they show promise of being able to measure distance to Earth within 1 cm. Optical clocks are now so accurate they can use the to measure an altitude of only 33 cm.[197]

---

[193] Ibid.
[194] http://go.glennborchardt.com/Roberts
[195] Borchardt, 2011, Einstein's most important philosophical error.
[196] http://go.glennborchardt.com/H-K1972
[197] Chou and others, 2010, Optical Clocks and Relativity.

An amusing implication is that your head is older than your feet.[198] There is now no question that clocks at some distance above Earth give different readings than those at sea level. The reason for this is often given as an effect of differing "gravitational potential," which indeed it is. What bugs the rest of us is that regressive physicists have no idea of what microcosm actually causes that. Like the rest of relativity and Newton's gravitational attraction, it is all just math. No physical cause—the collision of one microcosm with another—is presented as part of those theories. Without aether, there is no "there" there. A clock existing in empty space all by itself measures nothing. That is one reason I claim that the guiding philosophy of relativity is today's "scientific world view": systems philosophy (Table 5). Remember, that same philosophy has the entire universe exploding out of nothing—a finite "system" without a mechanical cause. No wonder there are 8,000 critics of relativity.[199] Without aether, relativity can never have a physical explanation. Adding to the confusion is the misnomer, "time dilation," which must have been invented by some reporter. That term, apparently never used by Einstein, did not even become popular until 1945. H-K never used it, preferring to use "relativistic time difference" instead. As mentioned throughout this book, time is motion. Only objects can dilate, so philosophically, you can forget about that impossibility. We need a better way to interpret the H-K results.

---

[198] Ost, 2010, Head older than feet.
[199] de Climont, 2016, ibid.

Table 7 Unpublished original test results and the Hafele and Keating alterations obtained by Kelly in 2000 (nanoseconds).[200]

Clock No.	Eastward Test Results	Eastward First Change	Eastward Second Change	Westward Test Results	Westward First Change	Westward Second Change
120	-196	-52	-57	413	240	277
361	-54	-110	-74	-44	74	284
408	166	3	-55	101	209	266
447	-97	-56	-51	26	116	266
Averages	-45	-54	-59	124	160	273

Note that the H-K experiment measured two different kinds of motion:

Table 8 Gravitational and kinematic contributions to the H-K results.[201]

	gravitational (general relativity)	kinematic (special relativity)	total	nanoseconds gained, measured	difference
eastward	+144 ±14	-184 ±18	-40 ±23	-59 ±10	0.76 σ
westward	+179 ±18	+96 ±10	+275 ±21	+273 ±7	0.09 σ

We already covered the "gravitational redshift" in our analysis of the Pound and Rebka experiment (Chapter 15.5) which has been considered a confirmation of General Relativity Theory. However, in lieu of "time dilation, we speculated that the velocity of light from a cosmic body increases with distance.

---

[200] Kelly, ibid, Table 2.
[201] http://go.glennborchardt.com/H-K1972

This is because aether pressure increases with distance from such a body per Aether Deceleration Theory. Aether particles permeate all matter. It is no more possible to prevent aether penetration than it is possible to prevent gravitation. That means a clock raised to a high altitude will experience an increase in supermicrocosmic impacts from aether. These, in turn, transfer some of their motion to the submicrocosms within the clock. Generally, a clock, being an instrument for measuring motion, will register an increase in internal motion as an increase in the number of ticks per second, which is measured as a time gain (Table 8). This is yet another confirmation of the existence of aether. Note that both the eastward and westward clocks had about the same time gains due to the gravitational effect. That is because both flights were at about the same altitude and aether pressure is the same at that altitude all around Earth. In essence, this part of the H-K experiment confirmed the existence of the "gravitational redshift" and provided support for Aether Deceleration Theory.

The second part of the H-K experiment involved kinematics and Special Relativity Theory. Apropos to special relativity, kinematics is "the branch of mechanics that deals with pure motion, without reference to the masses or forces involved in it."[202] In other words, kinematics purposely avoids using *causality*, the collision of one thing with another per Newton's Second Law of Motion. Neomechanics, of course, is all about just that. At first, I tried to understand the eastward data by applying neomechanics, but ran into a big contradiction: A moving clock is supposed to lose time (a decrease in submicrocosmic motion) and increase mass (an increase in submicrocosmic motion). Back to kinematics. It turns out the kinematic part of the experiment is not due to Einstein, but to

---

[202] http://go.glennborchardt.com/kinematic

Lorentz, who devised the correction factor $(1/\sqrt{1-(\frac{v^2}{c^2})})$ that Einstein misinterpreted in much of Special Relativity Theory.

Be it known that the Lorentz equation does not predict the magical dilation of time, slowing of clocks, increases in mass, and lengthening as a function of velocity. It simply and correctly explains the problem that arises when measurements rely on sending a signal through a medium (Figure 37).

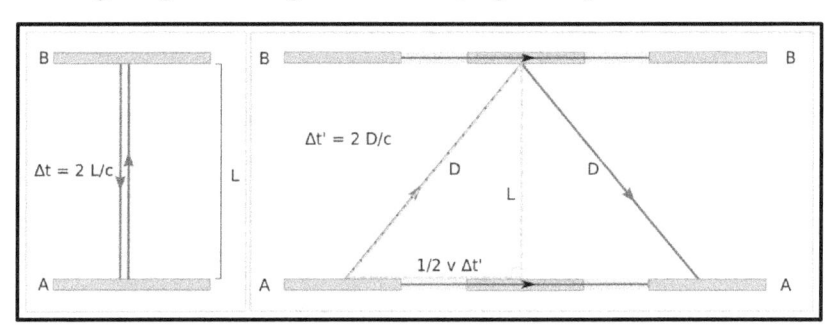

Figure 37 How signal transmission through a medium affects *measurements* of objects in motion. The path from point A to point B is simply lengthened when B is in motion, producing a measured increase in distance and the time required for the transmission to occur).[203] Credit: Sacamol.[204]

If *c* was infinite, the Lorentz correction factor would be 1.0 and the measurement of time for the stationary and moving clock would be identical. It makes no difference what the medium happens to be. Sound in air will do the same, but instead of *c*, the wave motion will occur at about 343 m/s. If you wish to know the true clock rate, all you have to do is divide your measurement by the correction factor. Remember, all this simply involves *measurement*. It has nothing to do with any changes undergone by the clock itself. It just keeps on ticking at its regular rate

---

[203] http://go.glennborchardt.com/TimeDilation
[204] (Own work), CC BY-SA 4.0,
    https://commons.wikimedia.org/w/index.php?curid=48778704

[unless you change its altitude (Table 9) or accelerate it (Table 10)]. That is analogous to the whistle on a train leaving the station. The train's motion through the air will increase the wavelength from that whistle, but the whistle on the train will have its usual frequency.

Table 9 Phenomena considered the result of "gravitational time dilation" (stationary clocks at high altitude).

	RELATIVITY INTERPRETATION
1	Cyclic frequency of clock increases
2	Period of clock decreases
3	Clocks have more ticks per second
4	Clock records more time
5	Clock gains time
6	Clock is fast
7	It takes less time for the same result
8	"Time would be gained"
9	GPS clocks run faster
10	Time gained is a function of altitude
	NEOMECHANICS INTERPRETATION
11	Clock has increased supermicrocosmic impacts
12	Clock absorbs increased motion from supermicrocosms (aether particles)
13	Time gain is a function of aether pressure, which increases with altitude
14	Prediction: Submicrocosmic motion of clock increases
15	Prediction: Mass of clock increases

Table 10 Phenomena considered the result of "kinematic time dilation" (moving clocks).

1	Cyclic frequency of clock decreases
2	Period of clock increases
3	Clock has fewer ticks per second
4	Moving clock records less time
5	It takes more time for the same result
6	Clock is slow
7	Clock loses time
8	"Time would be lost"
9	Time lost is a function of velocity
10	A moving clock ticks slower than a stationary clock when going east
11	A moving clock ticks faster than a stationary clock when going west
12	As in the Sagnac experiment, a clock going east around Earth undergoes a redshift, since it is moving away from the stationary clock in DC.
13	As in the Sagnac experiment, a clock going west around Earth undergoes a blueshift, since it is moving toward the stationary clock in DC.

The H-K experiment clearly indicated that travel westward produced the opposite of what was predicted by Einstein for high-velocity travel through empty space (compare Table 8 and Table 10). All moving clocks are supposed to slow down, but those going westward sped up instead. H-K considered the westward discrepancy to be a reference-frame problem. Schlegel showed the east-west variation to be a result of the Sagnac effect.[205] A clock leaving its base station eastward would take more time to return because the counter-clockwise rotation of Earth keeps moving the base station away from it—a redshift (long gray line in Figure 31). A clock leaving its base station westward would take less time to return because the rotation of

---

[205] Schlegel, 1973, Flying Clocks and the Sagnac Effect.

Earth keeps moving the base station toward it—a blueshift (short black line). The eastward clock would measure fewer ticks (Table 10) and the westward clock would measure more ticks than the stationary clock simply because the distance traveled east was more than the distance traveled west. Unfortunately, as an aether denier, Schlegel considered the results to be yet another proof of Einstein's particle theory of light. Remember, however, that Sagnac proved that light was a wave in the aether and that it clearly was not a *classical* particle. Again, to believe otherwise, as Schlegel and other regressives do, one must accept Einstein's outrageous ad hocs (Table 6).

Next, I will highlight another, less contentious experiment to show that "time dilation" is merely an artefact of measurement. This more recent experiment involves a special laboratory setup in which the "clock" is a single ion: Al^+ (Figure 38). You can see clearly that the measurement of time is correlated with the velocity of the moving clock. The curve shown is derived from the Lorentz correction factor. The data points fit the curve within measurement error. These regressives still do not realize so-called "time dilation" actually is due to transmission delays in the aether per Figure 37.

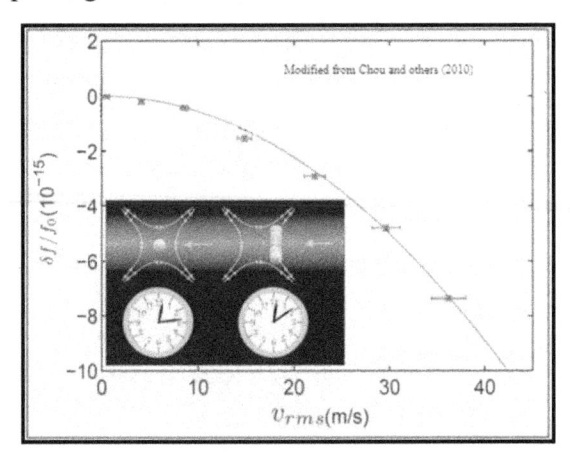

Figure 38 Kinematic time difference experiment. The frequency (f_o) decreases as a function of velocity (v_{rms}) when the clock is in motion.

Here is the authors' description of the above figure:

> *Relativistic time dilation at familiar speeds (10 m/s = 36 km/hour)... As the Al^+ ion in one of the twin clocks is displaced from the null of the confining RF quadrupole field (white field lines), it undergoes harmonic motion and experiences relativistic time dilation. In the experiments, the motion is approximately perpendicular to the probe laser beam (indicated by the faint shading). The Al^+ ion clock in motion advances at a rate that is slower than its rate at rest. In the figure, the fractional frequency difference between the moving clock and the stationary clock is plotted versus the velocity ($v_{ms} = \sqrt{(v^2)}$) (rms, root mean square) of the moving clock. The solid curve represents the theoretical prediction. The vertical error bars represent statistical uncertainties, and the horizontal ones cover the spread of measured velocities at the applied electric fields.*[206]

15.7 Did LIGO prove there are gravitational waves? (2016)

No. Instead, LIGO destroyed both gravitational attraction and aether denial as viable theories. Ever since gravitational waves were hypothesized by Einstein, regressives tried to get data proving him right. Admittedly, it gets a bit confusing because, according to General Relativity Theory, gravitation is supposedly caused by the curvature of space-time. In this neomechanical analysis, I will assume the signals obtained by the Laser Interferometer Gravitational-Wave Observatory (LIGO) (Figure 39) are valid.

---

[206] Chou and others, 2010, Optical clocks and relativity.

How LIGO works

Two special facilities in the U.S. — one in Louisiana, the other in Washington state — "listen" for gravitational energy that can indicate when and where cataclysmic events occurred in space. Here's how these laser interferometer gravitational wave observatories, known as LIGO, identify a distortion of space and detect those waves.

1 Laser

|—— 2.5 miles ——|

Interferometer

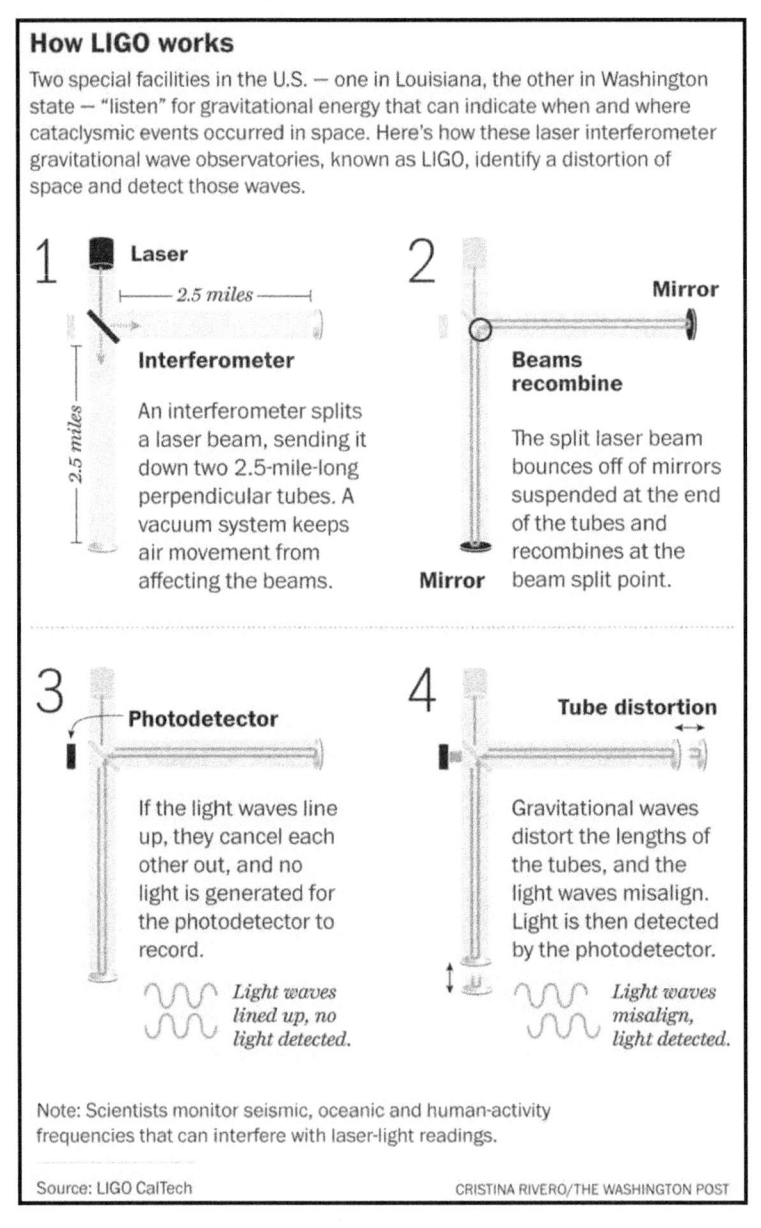

An interferometer splits a laser beam, sending it down two 2.5-mile-long perpendicular tubes. A vacuum system keeps air movement from affecting the beams.

2 Mirror

Beams recombine

The split laser beam bounces off of mirrors suspended at the end of the tubes and recombines at the beam split point.

Mirror

3 Photodetector

If the light waves line up, they cancel each other out, and no light is generated for the photodetector to record.

Light waves lined up, no light detected.

4 Tube distortion

Gravitational waves distort the lengths of the tubes, and the light waves misalign. Light is then detected by the photodetector.

Light waves misalign, light detected.

Note: Scientists monitor seismic, oceanic and human-activity frequencies that can interfere with laser-light readings.

Source: LIGO CalTech   CRISTINA RIVERO/THE WASHINGTON POST

Figure 39 Experimental setup for LIGO. Credit: LIGO CalTech.

First, let me quote from what I wrote about it in 2007:

> *Current government-supported work involves the search for gravitational waves, the general idea being to detect the results of explosions or collapses of celestial bodies. The*

success of the project would put the kibosh on the attraction hypothesis. Systems philosophy would shudder, but probably would regain composure by interpreting the data as yet another "proof" of Einstein's "space curvature" as the "mechanism" of gravitation. Of course, it really would prove nothing more than that space is not empty and that it contains a material medium capable of transmitting motion over great distances.[207]

Second, let me mention this is just another Einsteinism: "correct prediction, wrong reason." Although the much-ballyhooed gravitational-wave press conference totally ignored it, the actual discovery of gravitational waves would mean *the attraction hypothesis is dead*—only a push hypothesis would make sense. The shock wave from a huge collision a billion light years away certainly did not arrive here because we are especially attractive. It is no different from the sound we hear when the explosions from far-away fireworks produce changes in air pressure that travel to our ears. If those explosions were close enough and large enough, we might consider them "gravitational waves" only because of their tendency to knock us over. LIGO did not measure gravitational waves, but *shock waves in the aether*.

Third, we can now give up the idea that space is perfectly empty. What regressives call "space" or "space-time" is only imaginary and has no properties that would allow wave motion through it, and none is claimed. The experiment confirms *aether must exist*. The press reports stating the waves traveled through perfectly empty space via compression and decompression are ludicrous. That could never happen. All wave motion requires a medium—there has to be something to compress and decompress. That is why Einstein's corpuscular theory of light is

---

[207] Borchardt, 2007, The Scientific Worldview, p. 190.

equally ludicrous. Wave motion without a medium is like having water waves without water.

Fourth, the experiment did add one bit of valuable information: the speed of the shock wave and the speed of light are identical. Again, that is because they both use the same medium: aether, which can only transmit wave motion at a maximum velocity of c. This was confirmed by more recent LIGO measurement 170817, which was the first to use triangulation (WA, LA, Italy) to determine the exact location of the cosmic bodies that produced the shock wave. The signals from the LIGO detectors and from electromagnetic detectors pointed at that location arrived at the same time, with a precision of $\pm 10^{-15} c$.[208]

Fifth, the correct theory of gravitation involves aether particle deceleration as described in Chapter 16.3. As mentioned, the theory states aether pressure tends to be highest away from baryonic matter—just the opposite of atmospheric pressure. The "halo" of reduced aether pressure around all baryonic matter then acts like a vacuum, causing errant material objects to be pushed toward massive objects. Thus, gravitation acts locally, in the same way a helium balloon is pushed upward in the atmosphere due to local pressure differences. Unlike other push theories, I do not hypothesize corpuscles or gravitons rushing toward objects from far away at superluminal velocities. The highly active aether particles are already there, surrounding and permeating all baryonic matter. Actually, the theory has much in common with the much-ignored push theory proposed by Newton three centuries years ago. The key feature of both theories is the presence of aether, which has just been inadvertently confirmed by the LIGO experiment.

---

[208] Anon, 2017, GW170817, Accessed 20171023
http://go.glennborchardt.com/LIGO

Chapter 16
Progressive Physics

It should be clear by now that the little side trip taken by physics since 1905 must be abandoned forthwith. "The Ten Assumptions of Science," univironmental determinism, and neomechanics form the foundation for its replacement, progressive physics, which is essentially a continuation of classical mechanics with the inclusion of the assumption of *infinity*. Once freed from the nonsense engendered by indeterministic interpretations of relativity, the possibilities for physics and cosmology become endless. *Infinity* forces us to put aether back into our worldview, with its power to answer basic questions with physical causes where only mystery, immaterialism, or pure math prevailed before. Here are a few of those questions.

16.1 Why can there be no matter without motion?

Matter (microcosms) requires motion because it must be able to withstand the impacts from external supermicrocosms to exist. As an example, a balloon must be filled with a gas (submicrocosms in motion) to withstand the impacts of the surrounding atmosphere. If these submicrocosms were not in motion, they would be compliant and would be compressed together so much the balloon would collapse and cease to exist as an inflated balloon. In reality, all submicrocosms are in motion, even if their motions may appear insignificant with respect to the macrocosm. In another example, remember all microcosms have mass, which is defined as the resistance to acceleration. This resistance is provided by the contents of the microcosm, submicrocosms that impact the insides of the microcosmic wall, giving it shape as well. These impacts have momenta, P=mv, without which they could not counteract the momenta of the supermicrocosms colliding with the outside walls of the microcosm. In essence, microcosms require univironmental interactions for them to form and to exist. Remember, microcosms are assemblages of their various submicrocosms that came together via convergence and eventually will come apart via divergence per *complementarity.* In essence, a microcosm without submicrocosms could not exist—it would consist of nothing at all.

The necessity for submicrocosms to be in constant motion is what makes Finite Particle Theory hopeless speculation. Neomechanics assumes matter always contains other matter in motion—a proposition that has never been falsified. In this regard, the neomechanical explanation of the $E=mc^2$ equation maintains that the motion of submicrocosms is transmitted across the microcosmic wall via collisions with supermicrocosms (Figure 24). When the supermicrocosms involved are aether particles, that motion is transmitted through the aether medium at the speed of light. If the supermicrocosms involved were the nitrogen and oxygen molecules of the atmosphere, that motion is transmitted through the air medium at the speed of sound: 343 m/s. The $E=mc^2$ calculation applies to all microcosms regardless of size or velocity whenever the transfer of motion involves aether. This is because all baryonic microcosms are bathed in and permeated by aether. All collisions between microcosms involve aether particles although many of those aether collisions are insignificant. That is because all microcosms, including aether, have temperature, in which they are always either gaining heat motion or losing heat motion (Figure 24). This becomes most obvious when we measure infrared radiation (which travels at c). We can use a far-infrared camera to detect animals in the dark (Figure 25).

The hypothesized finite particles of the idealist do not have submicrocosms, so finite particles would have to be bizarre exceptions to the phenomena described by the $E=mc^2$ equation. Remember also that perfectly solid matter cannot exist—it is simply the ideal end member of the matter-space continuum. These end members are only ideas. They help us to understand real things—microcosms—that *do* exist. On the other hand, indeterminists sometimes suggest ideal solid matter and perfectly empty space are real possibilities. That conjecture is based on *absolutism*, the indeterministic opposite of **relativism**. Here is another incidence in which philosophy can enlighten physics. Although ideals such as perfectly empty space and perfectly solid matter do not really exist, an idealist always can assert they might be found in the future. This is where philosophy comes in. One has a clear choice between **relativism** and *absolutism*. Moreover, as I mention often, fundamental assumptions like these can never be proven. That is why we call them fundamental. Debates about fundamental assumptions are endless. All we can do is assume one or the other and go on with our work. What will that work be? As determinists, we assume **relativism** and therefore consider Finite Particle Theory to be wasted effort. The required "solid matter" is only an idealization. It cannot exist.

Infinite Universe Theory is the key to understanding why matter cannot exist without being in motion. That is because microcosms must contain submicrocosms whose impacts against the microcosmic boundary are required to resist the encroachment of the supermicrocosms in the macrocosm. On the other hand, without those incursions from the macrocosm, the submicrocosms constituting the microcosm would diverge from one another. Violating **conservation**, cosmogonists could imagine they would disappear into the nothingness of their equally imagined "empty" space.

16.2 What is aether?

Before the regression in the 20th century, aether was well accepted as theoretically necessary for light transmission as a wave.[209] The misinterpretation of the Michelson-Morley experiment (Chapter 14.1) led to its abandonment by Einstein and a return to the corpuscular theory of light long ago favored by Newton and Lavoisier. Although aether is responsible for many effects, its particle size is so small there currently are great difficulties in detection. This is not surprising for, according to *infinity*, there always will be microcosms so small they will be undetectable. This flies in the face of the operationalist's claim that undetected particles do not exist—a view consistent with the indeterministic assumption of *finity*. Of course, many particles were thought theoretically necessary long before having been discovered. Our ancestors could not see air either, but its effects were well known, demanding some physical explanation. Aether is like that.

Note that aether denial is a primary characteristic of regressive physics even though the Sagnac Experiment conclusively demonstrated aether's existence. As we saw, the only way relativists could get around that was to support Einstein's special pleading that gave magical properties to his light corpuscle. Nonetheless, the true nature of aether always has been speculative. I mentioned various applications of aether as necessary for understanding the Infinite Universe. Here I will summarize my own speculations as to the properties of aether. These may be magnitudes off, but from them you might appreciate the enormity of the problem and why speculation is needed. I developed these from conventional sources by making

---

[209] Whittaker, 1910 [1951], A history of the theories of aether and electricity: The classical theories; Whittaker, 1953, A history of the theories of aether and electricity: The modern theories; Shaw, 2016, Reconsidering Maxwell's aether.

certain secondary assumptions that probably will change in the future. For instance, I assumed Planck's[210] constant reflects the "smallest unit of motion" that we can detect, which I also assumed could only be the result of the collisions of aether particles[211] with baryonic matter. By using those assumptions, I was able to use Planck's equation to calculate what I presently think may be the properties of aether particles and the aether medium (Appendix). Table 11 is a summary of the results.

There is a prominent question bedeviling aether theorists for more than a century. Why does light motion occur as T waves? For physics beginners: An L wave compresses and decompresses the medium in the direction of travel. A T wave compresses and decompresses the medium in all directions perpendicular to the direction of travel. The fact we can polarize light proves that it is a T wave, despite some desperation on the part of other aether theorists. Gases and liquids generally have L waves.[212] Only solids have T waves (in addition to L waves). This makes aether a strange beast, with none of the transmission properties of gases and liquids and only one of the properties of solids. Of course, we should not necessarily expect aether to behave exactly like baryonic matter.

On the other hand, ***relativism*** teaches us aether particles must have some of the characteristics of other microcosms. As we showed in "Universal Cycle Theory," one of the most prevalent structural forms in the universe is the vortex. Obvious examples

---

[210] http://go.glennborchardt.com/Planck
[211] Many of these aether collisions formerly were attributed to photons, which are imaginary and therefore cannot be part of the aether medium. Photons are defined as the corpuscles that travel through empty space from galaxy to eyeball. Local aether particles also contact eyeballs, but, as in all wave motion, they are seldom the same ones emitted from a source. Although many of the properties attributed to photons and aether particles are identical, travelling from galaxy to eyeball is not.
[212] One exception: Waves on the surface of liquids are T waves.

are the solar system, Saturn, and the Milky Way. Back in 2009, I used vortex theory to speculate about the structure of the electron and positron.[213]

Table 11 Summary of speculative calculations on the properties of aether particles and the aether medium (see Appendix for details about how these were calculated from Planck's constant and the known electron mass).

Aether Particles
Mass=10^{-47} g
Diameter=3×10^{-19} cm
Volume=10^{-57} cm^3
Density=10^{10} g/cm^3
Shape=vortex disc (Figure 40)
Particle velocity=variable, near zero to 4.5 X 10^8 m/s
Composition=aether$_{-2}$ particles
Number in an electron=10^{20}
Aether Medium
Velocity=variable, approaching 3 X 10^8 m/s
Primary type of wave=T wave

---

[213] Borchardt, 2009, The physical meaning of E=mc^2.

Figure 40 The Sombrero Galaxy (M104). Does an aether particle look like this vortex disc? Credit: HST/NASA/ESA.

Thus, according to "Universal Cycle Theory," it is likely that aether$_{-1}$ particles are vortices formed from aether$_{-2}$ particles. Vortices form disc-like shapes as rotation rates increase (Figure 40). If this speculation is correct, then it appears likely a medium filled with these aether discs would seldom produce head-on compression and rarefaction in the direction of travel. Head-on collisions between disc edges would be rare, giving way almost entirely to wave motion perpendicular to the direction of travel. Per our observations, L waves[214] are insignificant, while T waves[215] dominate.

Mainstream physics views use of the "a" word (aether) as little short of treason. Trolls with an the inability to think on their own and too much time on their hands roam the Internet, ready to reprimand aether proponents as "cranks" or "crackpots." But with regard to aether denial, Robert Louis Kemp wrote this:

---

[214] http://go.glennborchardt.com/Lwavegif
[215] http://go.glennborchardt.com/Twavegif

> *Various models of the aether are being published in current scientific journals under different names: Quintessence, Higgs Field, Vacuum Expectation Value Energy, Zero Point Energy, Weakly Interacting Massive Particles (WIMPs), and Ground State Energy. All are Aether Theories at their core, each with their own twist, but Aether theories never-the-less!*[216]

The Michelson-Morley (1887)[217] experiment failed to measure the full 30-km/s velocity of Earth's revolution about the Sun because it was performed under the indeterministic assumption ether was fixed. In that case, Earth would face an "ether wind" as it moved through the stationary ether, just like the wind in your face when you run on a calm day. Their measurement was much less than the 30 km/s they expected although, according to Bryant,[218] it had a less than 0.1% chance of being the null result claimed by regressive physicists ever since. They got a low result because aether is entrained, much like Earth's atmosphere. Trying to detect the full 30 km/s at low altitude as they did, would be like trying to measure the jet stream in your backyard. The experiment has been repeated many times, with no hint of a null result. Because aether is entrained per Aether Deceleration Theory, the results are a function of altitude (Figure 41). To get anywhere near the 30 km/s velocity, the experiment would have to be repeated on the space shuttle or on a satellite—a nice job for NASA (don't hold your breath).

---

[216] Kemp, 2012, Super Principia Mathematica.
[217] Michelson and Morley, 1887, On the relative motion of the earth and the luminiferous ether.
[218] Bryant, 2016, Disruptive.

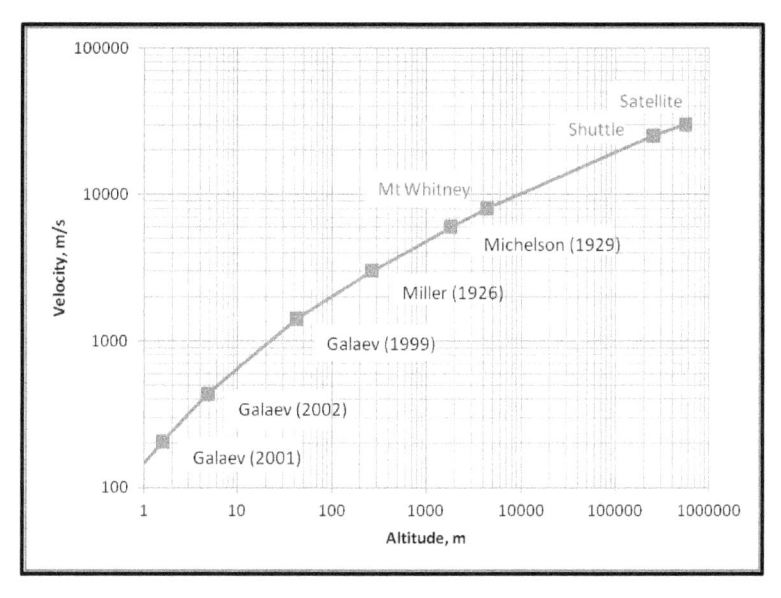

Figure 41 Interferometer measurements of Earth's velocity around the Sun as determined at various altitudes above mean sea level. The three data points at high altitude are projections and are yet to be performed. The other data are from Galaev, who seems to be the first to show this relationship.[219]

16.3 What causes gravitation?

Theories about the physical cause of gravitation are nonexistent in regressive physics, which does not hypothesize a physical cause. The "attraction" theory attributed to Newton and the "curved space-time" theory attributed to Einstein clearly do not propose physical causes. No one, including Newton, has ever explained how attraction works. Are there little unseen microcosms with hooks involved? Is the "pull" of gravitation acausal? Einstein's curved space-time was empty until his recantation in 1920. Even then, there was not supposed to be anything "physical" about it that could produce curved space-time or anything else. Attraction and curved space-time were to continue as the imaginings of those plagued by aether denial.

---

[219] Galaev, 2002, The measuring of ether-drift velocity.

To demand we explain gravitation as the result of a physical cause is to demand we approach the phenomenon from the mechanical point of view: things hitting things. Newton's equation was fine, but equations do not do anything. They are descriptions of nature—they do not run the universe. The universe does that fine by itself, thank you. As I mentioned numerous times, the "physical" in causality is based on Newton's Second Law of Motion in which a cause is defined as the collision of one thing with another. This means, of course, that gravitation must be a push, not a pull. Push ideas have been around as long as the pull idea—the equation fits either.

One of the most popular is called the Le Sage theory,[220] which was proposed in 1690 by Nicolas Fatio de Duillier,[221] a friend of Newton, and reproposed by Georges-Louis Le Sage in 1748. I highlighted it in "The Scientific Worldview" because it was the best push theory available at the time.[222] Le Sage hypothesized corpuscles that traveled at high velocity from all directions, permeated baryonic matter, and produced shadows of reduced pressure between objects, causing gravitation. Still, there are serious problems with it. Unlike light, gravitation has no aberration. Aberration is reflected in the 8-minute delay necessary for light waves to travel from Sun to Earth. Gravitation exhibits no such delay—calculations of planetary orbits must use their current positions, not some positions they once had. Van Flandern famously calculated that the velocity of those hypothesized gravitational corpuscles therefore must be 20 billion times the velocity of light![223] Such fantastic velocities would require a medium other than aether. Unless you wish to include that ad hoc, Van Flandern's simple calculation falsifies

---

[220] Edwards, 2002, Pushing gravity.
[221] http://go.glennborchardt.com/Duillier
[222] Borchardt, 2007, The Scientific Worldview, p. 189.
[223] Van Flandern, 1998, The speed of gravity.

not only the Le Sage theory, but also all theories hypothesizing particles or waves that must travel extremely long distances before doing their pushing job.

Long-Range vs. Short-Range Travel

That these long-range gravitation theories are still being taken seriously in the reformist community[224] implies a bit more discussion is needed here. In radiation theory, range[225] refers to the distance a microcosm travels before colliding with another microcosm. Alpha particles,[226] for example, exhibit short-range travel. They can be stopped by only a few sheets of paper. The concept of long-range particle travel reached its climax with the invention of the photon, which was supposed to travel uninterrupted through empty space from galaxy to Earth per Einstein's Untired Light Theory (Chapter 7). As seen in the discussions above, we no longer believe that amount of long-range particle travel is likely—at least not for light, which is a wave and not a particle.

Media generally exhibit short-range particle travel. For example, the inter-particle travel of the nitrogen molecules in air may be up to 1.5 times the velocity of sound, but the distances are very short. Nonetheless, the waves produced in that medium can travel very long distances. Light behaves the same way. I speculate that the inter-particle travel of aether particles is analogous, being up to 1.5 times the velocity of light, and that those distances are very short as well.

---

[224] Schroeder, 2006, The universe is otherwise; Shaw, 2012, The Cause of Gravity.
[225] http://go.glennborchardt.com/range
[226] Positively charged particle identical to the nucleus of a helium-4 atom (e.g., 2 protons and 2 neutrons).

Aether Deceleration Theory

As seen in all the previous discussions, infinite divisibility and *interconnection* implies the existence of intervening microcosms performing many of the functions generally attributed to aether. As regressive physics has demonstrated so unabashedly, without aether, it is impossible to understand light, gravitation, and, as we shall see later, the formation of baryonic matter itself (Chapter 16.4). Granted, we may never develop methods for observing aether particles directly, probably because all our detectors are made of aether complexes. Aether is theoretically necessary nonetheless. That is why Sagnac was able to show light was a wave and not a particle.

As we speculated in our previous paper on gravitation,[227] the relative activity or pressure of free aether appears to be dependent on the absence of complexed aether—ordinary matter. For instance, we interpreted the gravitational redshift documented in the Pound-Rebka experiment[228] as the result of increases in velocity as light travels through a medium that becomes more active with distance from massive bodies. That was one of the first clues as to the nature of gravitation. If aether particles also are responsible for gravitation, then the decreased pressure near massive bodies would act like a vacuum. It would be the reverse of the reduction in atmospheric pressure that occurs with increases in elevation. For Earth, high aether pressure at high elevations would cause matter to be pushed toward areas of low aether pressure at low elevations. Gravitation would be local. It would involve short-range particle travel, not long-range travel as proposed in previous pushing theories falsified by lack of aberration.

---

[227] Borchardt and Puetz, 2012, Neomechanical gravitation theory.
[228] Pound and Rebka, 1960, Apparent Weight of Photons.

Now for the critical question answered in this section: Why is aether pressure higher away from baryonic matter than near it? In the previous paper, we attributed this to particle-size separation due to vortex rotation. The main question left unanswered: If vortex rotation is required to initiate the pressure differences, then why do non-rotating microcosms still exhibit gravitation?

It turns out the answer is obvious and required by our initial assumption gravitation is a push and not a pull. It also is obvious from our insistence all causes are mechanical—the collision of one thing with another. Collisions that produce gravitation, the acceleration of ordinary matter, invariably, and most importantly, must cause the aether particles involved in those collisions to decelerate per Newton's Second Law of Motion (Figure 42).[229]

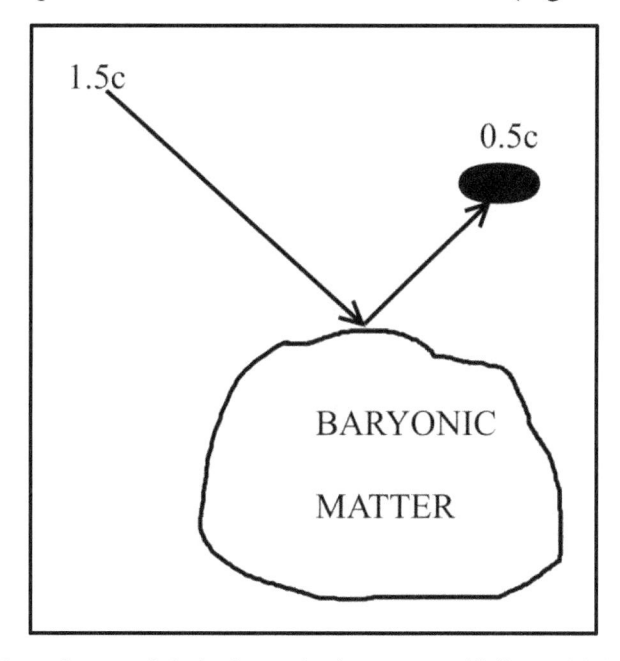

Figure 42 Aether particle losing velocity upon colliding with baryonic matter.

---

[229] A good example of such occurs when pendulums collide, with the collider stopping and the collidee accelerating as the motion is transferred from collider to collidee. See Figure 14.

This means, of course, that ordinary matter must be surrounded by what I like to call "decelerated, slowed, tired, dead, or spent aether particles," producing a halo in which the velocity of aether particles is diminished. In other words, any object accelerated by the push of gravity must leave behind very tired aether particles that did all the work described by Newton's $F = Gm_1m_2/r^2$ equation. Because of these collisions, the "graveyard" of these decelerated particles must extend in all directions, its effect diminishing inversely with distance per the r^2 term in the equation.

Because this theory of gravitation is such an important part of Infinite Universe Theory, let me repeat its essential elements. As explained in the neomechanics section, I assume all microcosms are continually bombarded by supermicrocosms that exist in the macrocosm. Aether particles are a large portion of that bombardment. Some of those collisions cause acceleration of the microcosm as a whole. In so doing, aether particles must lose some of their velocity. The deceleration appears among aether particles closest to the microcosm, producing a sort of "graveyard" of relatively densely packed, decelerated aether particles. The density of the aether medium would be an inverse function of its activity. Remember, that in gases, we measure activity as pressure. Thus if we cool a container of hot gas (a submicrocosmic deceleration), the molecules within will slow down and the pressure will drop. The upshot is that aether particle deceleration produces a Gravitational Pressure Gradient whose pressure *increases* with distance from every microcosm. This is because the distal portions of the macrocosm are increasingly populated with high-velocity aether particles that have yet to collide with the microcosm.

Again, this gradient is analogous to the atmospheric pressure gradient surrounding Earth—but in reverse. A helium balloon will rise in our atmosphere because the impacts of air molecules are greater from the high-pressure regions below than from the low-pressure regions above. A massive object will do just the opposite in Earth's gravitational field. It will be pushed toward Earth because the aether impacts are greater from above than below. After colliding with baryonic matter, an aether particle necessarily loses some of its motion while the baryonic matter gains motion. These decelerated aether particles tend to produce a sort of "aether vacuum" around all baryonic matter—a region in which aether velocities are diminished and aether pressure is low. Of course, the total absence of aether never occurs because aether particles are so numerous and so small. They permeate all ordinary matter, with the impacts producing gravitation in baryonic matter being infrequent and proportional to mass, which we define as the resistance to acceleration. In other words, aether flows through and around every larger microcosm, with the contact that produces gravitation being most evident for the densest microcosms. In the helium balloon example, nitrogen and oxygen molecules, being of greater mass than helium, tend to succumb to aether impacts more easily than does helium. Helium, in turn, tends to succumb more to impacts from nitrogen and oxygen than from impacts due to aether.

The upshot is that the physical cause of gravitation is due to aether pressure differences, which are local and ever proportional to the mass of the microcosm of concern, just as Newton said it was. The common view is that Newton denied knowing the physical cause of gravity (i.e., *hypotheses non fingo*), considering it to be an immaterial attraction. He later changed his mind, offering a push theory on page 325 in Query 21 of the second edition of Opticks,[230] written in 1718, a part of which I put below (Figure 43). As we mentioned in our earlier paper, Newton had a similar idea, although he got things backwards.

---

[230] Newton, 1718, Opticks.

> [325]
> Qu. 21. Is not this Medium much rarer within the denfe Bodies of the Sun, Stars, Planets and Comets, than in the empty celeftial Spaces between them? And in paffing from them to great diftances, doth it not grow denfer and denfer perpetually, and thereby caufe the gravity of thofe great Bodies towards one another, and of their parts towards the Bodies; every Body endeavouring to go from the denfer parts of the Médium towards the rarer? For if this Medium be rarer within the Sun's Body than at its Surface, and rarer there than at the hundredth part of an Inch from its Body, and rarer there than at the fiftieth part of an Inch from its Body, and rarer there than at the Orb of *Saturn*; I fee no reafon why the Increafe of denfity fhould ftop any where, and not rather be continued through all diftances from the Sun to *Saturn*, and beyond. And though this Increafe of denfity may at great diftances be exceeding flow, yet if the elaftick force of this Medium be exceeding great, it may fuffice to impel Bodies from the denfer parts of the Medium towards the rarer, with all that power which we call Gravity.

Figure 43 Newton's push theory of gravitation in which he hypothesized a universal medium with increasing distal density.[231]

He offered proximal density decreases instead of distal pressure increases as the cause of gravitation. We are certainly reassured to arrive at a similar conclusion independently, even if

---

[231] Newton, 1718, Optiks, p. 325.

three centuries late. Looks like those of us in progressive physics need to do a better job of searching the original sources.[232] Newton is usually blamed for his early indeterministic attraction theory, but the deterministic push theory he proposed has been swept under the rug by regressive historians. This illustrates an important pedagogical fact: We must be continually aware that, as in all battles, the results of philosophical and scientific struggles are always written by the victors.

Entrainment and the Aether Halo

A zone of reduced aether pressure exists around every baryonic object. Earth's atmosphere is analogous, except the pressures are reversed (Figure 44). Of particular importance is the fact that the atmosphere is entrained, that is, it moves along with Earth as if it were part of Earth itself. That is because the pressure at each point in the atmosphere is in continuous equilibrium with all adjacent points. Imagine, if you will, that the atmosphere consisted of solid metal, or of plastic just like a DVD or CD. Any rotation of the center of that disc would "instantaneously" appear as rotation of its outer edge. The aether halo produced by gravitation behaves the same way. That is why gravitation displays no aberration. It is as if the Moon were attached to a solid disc surrounding Earth (Figure 45). Every rotation of Earth is accompanied by an equivalent nearly instantaneous rotation of the Moon.

Remember, light from the Sun displays aberration because it takes 8 minutes for light to get here. That is because light is a wave in the aether. By the time that image of the Sun gets here, the Sun is no longer where its image says it is. For instance, if that image, taken 8 minutes ago, showed a star in the background

---

[232] Thanks to Duncan Shaw for the reference to Newton's long-forgotten push theory for the physical cause of gravitation.

near the Sun it would not be representative of the present. The rotation of Earth could be enough to put that star out of the picture altogether.

On the other hand, entrainment means Earth is, in effect, attached to the Sun's gravitational halo, and vice versa. This association, produced entirely by aether pressure relationships, remains relatively constant and continuous, almost as if the Earth and Sun, like the Earth and Moon, were on a solid disc. None of this conflicts with the attraction hypothesis—except for the physical cause. As mentioned, Newton's equation works whether gravitation is a pull or a push. The only difference between the two hypotheses is that one requires a belief in the magic of attraction and the other requires a belief in the reality of aether.

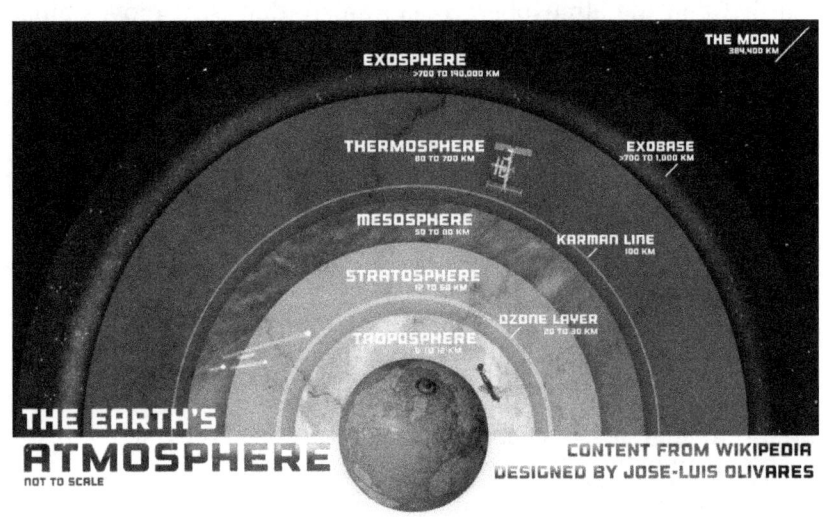

Figure 44 Earth's atmosphere.[233]

---

[233] http://go.glennborchardt.com/Atmosphere

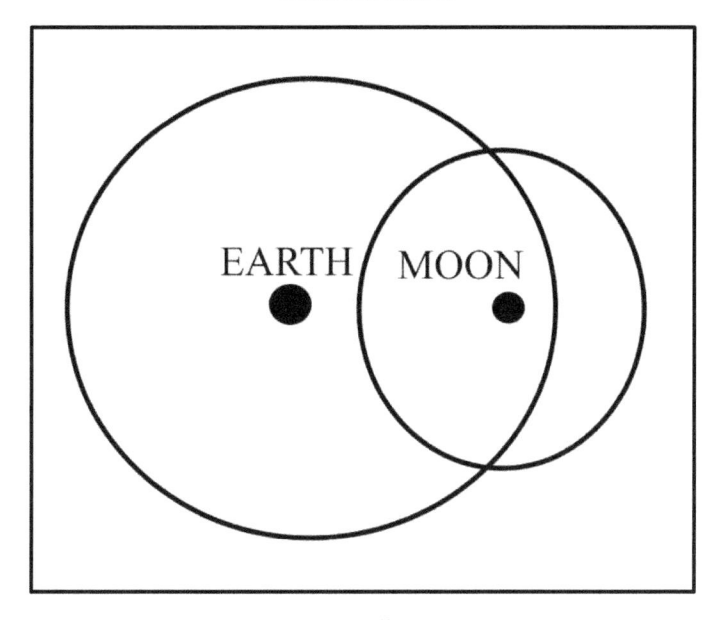

Figure 45 Earth-Moon gravitational discs.

16.4 Where does matter come from?

From somewhere else. That was my scientific answer to our four-year old daughter when she asked: Where did all this stuff come from? She was referring to the view of Marin County she could see from the backseat of our car. My answer was prescient because I was barely half way on my journey toward "The Scientific Worldview." It turns out only an Infinite Universe could exist. In the Infinite Universe, each thing is a combination of other things that converged temporarily from elsewhere. The "stuff" she was thinking of and the "matter" we are discussing were the same. Both words are abstractions for xyz portions of the universe. There is no "stuff" *per se* and no "matter" *per se*. There are only unique examples of each, with each example containing still other examples of matter ad infinitum.

Through *relativism* we assume no two microcosms, in this case aether particles, are identical. Identical particles would have no reason to form complexes. That is another reason Fundamental Particle Theory will never be successful. These particles imagined by idealists must be perfectly spherical and perfectly solid. If they were not perfectly spherical, they would betray the presence of some internal structure made up by submicrocosms. That would mean they were not fundamental. In that case, the *submicrocosms* would be "fundamental." While fundamental particle theorists must be forever bedeviled by that problem, we see the dissimilarities between aether particles as a requirement for the production of baryonic matter from aether particles. Indeed, this is in line with a major principle of Infinite Universe Theory: A perfect world cannot exist. Those infinitely unique imperfections in each microcosm not only allow it to exist for a time, but those imperfections are necessary for its coming into existence from other matter in the first place.

Although the rest of this explanation necessarily jumps into the middle of the infinite hierarchy, the process applies to aether particles and galaxy clusters alike. In intergalactic space, each aether particle has a unique momentum (i.e., P=mv, where m=mass and v=velocity), with the potential to transfer some or all of its motion to other aether particles (Figure 14). This process continues indefinitely just as it does among the nitrogen molecules that continually collide with each other in the atmosphere. Like those nitrogen molecules, aether particles are so similar that most of their interactions produce nothing new. The accelerating and decelerating goes on incessantly, allowing nitrogen molecules with particle velocities up to 515 m/s to "instantaneously" conduct sound waves at 343 m/s and for aether particles with velocities up to 450,000,000 m/s to conduct light waves at velocities up to 300,000,000 m/s.

However, per *relativism* no two microcosms can be identical or perfectly spherical. This is illustrated in Figure 46, which shows the irregular shapes and sizes common to microcosms. Remember, in neomechanics all microcosms are assumed to exist in irregular shapes because they contain submicrocosms.

Only the "fundamental particles" of the idealists are perfect spheres filled with solid matter—and they do not and cannot exist. Again, this is a critical point for Infinite Universe Theory. Without the variations and imperfections common to all microcosms, the Infinite Universe would not be possible. Without the imperfections generalized in Figure 46, we would not be here either. If all microcosms were identical, as implied by the atomists, there would be no reason for any of them to combine to form new microcosms. The watchword for nature is *vive la différence*.

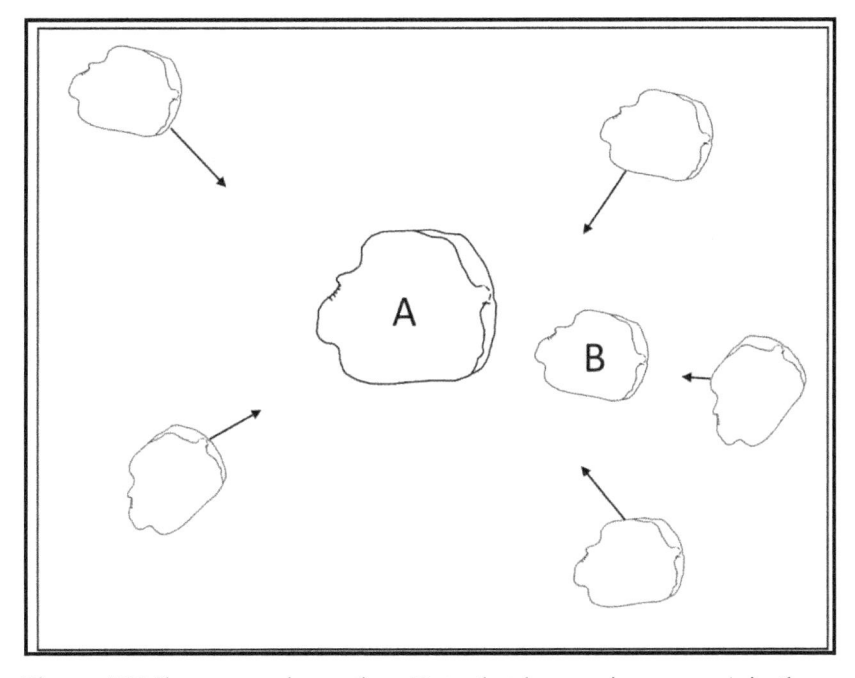

Figure 46 Microcosms in motion. Note that large microcosm A in the center shelters microcosm B from impacts from the left. Consequently, B will be pushed toward A, with the likelihood it might even end up rotating around A or combining with it.

As the figure shows, some microcosms are more massive and likely slower than others. Large, less active microcosms will be pushed together by the less massive, faster aether particles. This phenomenon occurs throughout the universe. Supermicrocosms in the macrocosm continually bombard every microcosm. It is why balloons or beach balls are pushed to the end of your swimming pool (Figure 47). It is why the world will push you around unless you push back. It is why wagon trains and musk oxen form a circle for protection. It is why corporations form mergers and workers form unions. The closer microcosms become, the more they shield each other from the impacts of the macrocosm.

Figure 47 Balloons pushed to the side of the pool illustrating the tendency for microcosms to be pushed together by impacts from the macrocosm.

Again, that is the key to the formation of baryonic matter from aether particles. The mutual shielding thus produces a complex, which, by definition is slower and clumsier than free aether particles—in the same way any social group or paradigm behaves in relation to those not so attached. In chemistry, such combinations are common. Na^+ and Cl^- ions, for instance, are forced to combine as NaCl crystals when the water evaporates and the Na^+ and Cl^- ions are pushed together, being confined to an increasingly restricted space. Vortices get involved because the largest aether complexes, being slow and heavy, tend to be pushed toward the center of any rotating cloud of such complexes. Still another way of visualizing this is to imagine the classic "round-up" that is necessary to stop a herd of cattle for the night. Cowboys on one side of the herd speed the animals on that side until the herd moves in a circle. An animal caught in the middle of the herd has no choice. It also must move in a circle. The pressures on both sides are equal, being part of a larger motion. All this simply results in the slowing down of aether particles to form aether complexes, becoming ever larger and forming what we know as baryonic matter. Some of the first baryonic microcosms probably were vortices such as electrons and positrons.

Now note the similarity between this formation of baryonic matter from aether and the cause of gravitation. In both cases, small microcosms collide with large ones, transferring some of their motion per Newton's Second Law of Motion (F=ma). In so doing, the colliding aether particles lose velocity, tending to remain near the large ones. These decelerated aether particles pile up, producing a zone of increased aether density and low aether activity near the aether complexes. In this regard, one could say, "matter begets matter." Gravitation is essentially the same process, with this reduction in aether activity appearing as a halo or zone of low aether pressure surrounding all baryonic matter. Thus, baryonic matter formation and gravitation are essentially the same process. Gravitation and aether complexification both result in the taming of high-velocity aether particles from the free field. In addition, any vortex rotation tends to convert much linear momentum into angular momentum. This further concentrates microcosms as submicrocosms within larger entities. The linear velocities of these microcosms decrease, while the submicrocosms within may continue to rotate at high velocities as in the "round-up" example.

Remember that because aether transmits mostly T-waves (side-to-side motions), I speculated aether particles were vortices (Figure 40). This also follows from our observation that matter in the cosmological realm tends to form vortices at all scales.[234] The vortex shape would make it especially easy for aether particles to form combinations (Figure 48). Variations in size would not be particularly important. I am unsure whether such stacking of aether vortices continues beyond the duplex stage. Nevertheless, remember my speculations from Planck's constant concluded a single electron might be a complex containing about 10^{20} aether particles (Table 11). This may seem like a lot, but an average snowflake is a complex containing about 10^{19} H_2O molecules.[235] No wonder no two electrons and no two snowflakes are identical. Such is the Infinite Universe.

Figure 48 Hypothetical aether particles showing the effects of vortex morphology. The two parallel vortices, each having exposure on one side, will receive fewer impacts than the others will. Credit: Sombrero galaxy images modified from NASA.

---

[234] Puetz and Borchardt, 2011, Universal Cycle Theory.
[235] Roach, 2007, "No Two Snowflakes the Same" Likely True.

As explained in our analysis of celestial microcosms, any increase in the rotation of a vortex produces an accretion of matter, while any decrease produces an excretion of matter.[236] As always, celestial microcosms come into being via convergence and go out of being via divergence. Rotations occur when microcosms collide tangentially, sideswiping each other to produce opposite spins. The rotation eventually stops after the vortex succumbs to friction produced by the macrocosm. This is another indication space is not perfectly empty. Without the supermicrocosms in space, vortices would rotate perpetually, which according to the Second Law of Thermodynamics cannot happen. Like all microcosms, vortices experience birth and death.

You might say: Well and good, you can explain the origin of baryonic matter from your hypothesized aether, but where did that aether come from? As always, per Infinite Universe Theory, the answer is "From somewhere else." This "passing of the buck" is an essential characteristic of the Infinite Universe as I surmised in my answer to our daughter long ago. As mentioned previously, in "Universal Cycle Theory" we handled this problem by assuming baryonic matter forms from $aether_{-1}$, that $aether_{-1}$ forms from $aether_{-2}$, and that $aether_{-2}$ forms from $aether_{-3}$ ad infinitum.[237] Again, this point is crucial for Infinite Universe Theory. Scale means nothing to the Infinite Universe. Without this infinite regression, the idealist's imagined "perfectly empty space," nothingness, and nonexistence would be possible. On the contrary, our very existence is obvious and provides support for Infinite Universe Theory.

---

[236] Puetz and Borchardt, 2011, Universal Cycle Theory.
[237] Ibid.

16.5 What is the cause of charge?

One of the greatest deficiencies of today's regressive physics is that there is no physical explanation for charge. In class, we are taught to put little minuses on symbols for electrons and pluses on symbols for protons. None of that tells us what is really going on other than the simple observation charge is a "property of matter" and opposite charges "attract" and like charges "repel." I suspect this mystery remains so prominent because charge is a univironmental phenomenon involving the much-maligned aether. In other words, charge, like so many properties of matter, is not a system property, but a relationship between the microcosm (e.g., electron) and its macrocosm (aether). Like so much of regressive physics, the physical cause of charge will remain a mystery until aether denial is finally abandoned. As you might imagine, aether denial has left a tremendous opportunity for dissidents to set things right. Much of that work is atrocious, but some of it is not. I am impressed by the work of Ionel Dinu, a young Romanian physicist, who uses fluid mechanics in his attempts to understand univironmental interactions involving aether.[238] Although I do not agree aether really is a liquid, he has produced a wonderful video demonstrating the likely physical mechanism for charge by using fluid mechanics:

Figure 49 Demonstration of univironmental "attraction" and repulsion by Ionel Dinu at http://go.glennborchardt.com/Dinucharge.

---

[238] Dinu, 2010, Rudiments of a theory of aether.

Again, the most important point illustrated here is that the fluid is critical for this illustration to work. In addition, the rotating microcosms do not rotate independently of their surrounding macrocosms. Like the entrained atmosphere and aetherosphere surrounding Earth, friction between microcosm and macrocosm is necessary for attraction and repulsion to occur. Here is a diagram of what is happening in Dinu's demonstration:

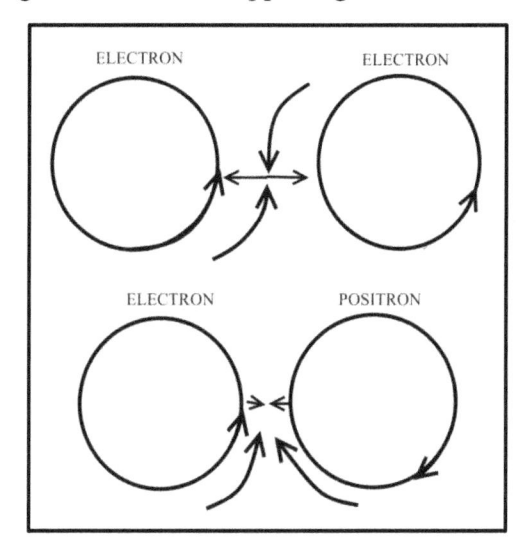

Figure 50 Diagram of how spin direction produces "attraction" and repulsion.

Electrons spinning in the counterclockwise direction will repel each other because they drag part of their surrounding macrocosms in opposite directions (Figure 50). The resulting collision between the two entrained macrocosms increases the pressure between them, producing a tendency to force the two electrons apart. Conversely, an electron spinning counterclockwise and a positron spinning in the clockwise direction will "attract" each other because they drag part of their surrounding macrocosms in the same direction. This produces a sort of vacuum between the two reflected in the Bernoulli principle,[239] which "states that an increase in the speed of a fluid occurs simultaneously with a decrease in pressure." And, as you know, "nature abhors a vacuum." The result is that the high pressure in the surrounding area pushes the electron and positron together.

If the rotating discs or the electrons and positrons were surrounded by perfectly empty space, they would neither "attract" nor repel each other. The upshot: Without aether, charge would be impossible. The details no doubt are more complicated than shown in the demonstration, but the lesson is clear—charge is not a "system" property, but a property of the system and its environment.

16.6 What is the cause of magnetism?

Aether denial has led to another of the great travesties produced by regressive physics: the missing physical explanation for magnetism. Einstein's gravitational and magnetic fields are immaterial—they are completely empty space. As I mentioned previously, there is no "there" there. Thus, when most folks are asked about it, they typically say "matter attracts matter" and "opposite magnetic poles attract each other." They only say this because it is what they were taught to recite in the classroom. Like their imagined journey into the hereafter, they are unlikely to think deeply about how this happens in a physical sense.

There is a magnet in our kitchen where we hang our knives. To the aether denier, there is nothing whatsoever responsible for the obvious continual bombardment that keeps the knives hanging there. That process might just as well be "magic." Again, despite the attraction hypothesis, Newton's three laws of motion describe only pushes. Any "pull" we can perform always involves a push. For instance, we must curve our fingers around

---

[239] http://go.glennborchardt.com/Bernoulli

a door handle to apply a pushing force toward ourselves. The suction of a vacuum cleaner is really a push. Obviously, the "physics" of magnetism must involve aether, the macrocosm missing from the myopic tendency for systems philosophy to ignore the environment.

Here is another of Ionel's videos, this time demonstrating how objects spinning in opposite directions "attract" one another and objects spinning in the same direction repel one another. In this simulation, he has prepared devices with two ends he can spin in opposite directions. By putting them in water, he can produce some of the movements akin to magnets. The point here is to show that the physical mechanism for magnetism must involve both the microcosm and its macrocosm.[240]

It is well known that magnetism involves the movement of electrons, but what is still unrecognized is the part played by the entrained aether surrounding each electron. Remember, gravitation is produced by aether particle deceleration when those particles collide with baryonic matter. Electrons are baryonic matter too, so F=ma collisions between aether particles and electrons must generate a "halo" of decelerated aether particles around each electron. Like gravitation, this entrained area of lowered aether pressure extends for a considerable distance. The effect follows the inverse square law and has the same form as the one discovered by Newton for gravitation:

$$F = k_e q_1 q_2 / r^2$$

Where:

F = force

k_e = Coulomb's constant

q_1 = charge on negative particle

---

[240] http://go.glennborchardt.com/Dinumagetism

q_2 = charge on positive particle

r = distance between particles

This relationship[241] was discovered by Charles Coulomb, a French physicist, about a century after Newton. The similarity between the two equations is no accident, since both phenomena are produced by the same mechanism: aether particle deceleration.

16.7 Why do satellites stay in orbit?

Satellites, like all microcosms, exist at a particular location for univironmental reasons: the matter in motion within and without. As mentioned, aether particle deceleration produces a halo of lowered aether pressure around all baryonic microcosms. F = ma collisions between aether particles and baryonic matter cause baryonic matter to be accelerated (per Newton's equation for gravitation) and aether particles to be decelerated—quite elementary. The slowing of aether particles shows up as a reduction in aether pressure. Again, per Newton's equation, the effect is an inverse function of distance.

As mentioned previously, in regressive physics the effect is considered a product of "gravitational attraction" that produces a "matterless gravitational field." In that model, a satellite is said to remain in orbit when its momentum produces enough centripetal force to counteract the "attractive force" produced by the cosmic body around which it revolves. In Aether Deceleration Theory, a satellite remains in orbit because its distal aether pressure is similar to its proximal aether pressure, with its momentum being sufficient to counteract some of the slight difference between the two. Aircraft flight through the atmosphere is analogous. Air pressure above the wings is similar to the air pressure below the wings. However, to counteract the

---

[241] http://go.glennborchardt.com/coulomb

push of gravity, air pressure above the wings must be less than air pressure below the wings for a plane to remain at constant altitude.

Satellites that always are at the same point above Earth are called geosynchronous.[242] They rotate at the same rate as Earth, making one revolution every 24 hours (Figure 51). This is yet another proof gravitation has no aberration and results, instead, from aether pressure differences within the halo produced by aether particle deceleration around Earth. The aether pressure differences are local, gradational, and constant, acting as if all satellites, including the Moon, are connected to Earth like the ends of the spokes on a wheel (Figure 51).

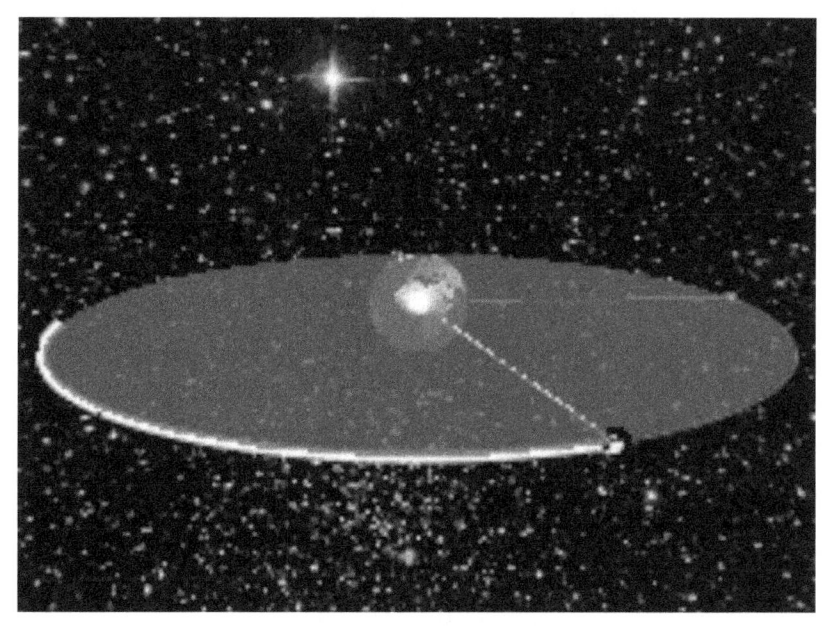

Figure 51 Geosynchronous satellites at about 36,000 km above Earth.[243] Credit: Lookang.[244]

---

[242] http://go.glennborchardt.com/geosysat
[243] http://go.glennborchardt.com/geosyngif accessed 20171120.

Another way of visualizing this is to imagine septillions of aether particles are entrained in the space around Earth, rotating with it as if they were part of a rotating metal disc. A satellite caught up in that disc would rotate along with it, with aether pressure being similar on the sides toward and away from Earth. Having greater momenta, some satellites revolve faster than others and some artificial satellites even have *retrograde*[245] motions (opposite the rotation of Earth). Artificial satellites are mostly prograde because it takes extra fuel to put satellites into retrograde orbits. Prograde launches get an inertial boost from Earth's rotation, while retrograde launches have to overcome it. Once in orbit, a satellite experiences drag due to the multitude of supermicrocosms in the macrocosm. For instance, those in low Earth orbit are slowed most significantly by the upper extents of Earth's baryonic atmosphere. Theoretically, aether should produce some deceleration or acceleration, particularly whenever entrained aether velocity does not match satellite velocity. This effect is unlikely to be measured because even satellites in high orbits probably would be more influenced by collisions with electrons and other atomic particles instead.

Note that a planet does not need to be rotating to have satellites. This is because an entrained aether halo forms around all baryonic matter due to aether deceleration. Again, slight aether pressure differences keep satellites in orbit around slow-rotating planets such as Venus (about one rotation per year). Because its aether halo is nearly stationary, both prograde and retrograde orbits initially require extra fuel for increased momenta if they are to rotate faster than Venus.

---

[244] By Lookang many thanks to author of original simulation = Francisco Esquembre author of Easy Java Simulation - Own work, CC BY-SA 3.0, https://commons.wikimedia.org/w/index.php?curid=15629545

[245] http://go.glennborchardt.com/retrosat

16.8 Why is there so much spookiness in quantum mechanics?

Quantum mechanics is fraught with numerous erroneous interpretations based on indeterministic assumptions underlying regressive physics. Much of this has been implied in previous discussion, but two concerns are primary: 1) violation of the Third Assumption of Science, **uncertainty** (It is impossible to know everything about anything, but it is possible to know more about anything) and 2) aether denial.

The first violation is known as the Copenhagen interpretation. It was an indeterministic response to the Heisenberg Uncertainty Principle, which proved the position and the velocity of a particle could not be known at the same time. Heisenberg destroyed classical mechanics, making its foundational assumption of *finity* untenable. Because the universe is infinite, any experiment anyone could perform always gave a different result when performed the second time. The truth was that neither the *exact* position nor the *exact* velocity of a particle could be determined. All measurements, whether of large or small microcosms, always had a plus or minus. The Copenhageners interpreted this, not as observer ignorance due to infinity, but as an inherent characteristic of a finite universe ruled by probability. In the classic tradition, they thought of an effect as being the result of a finite number of causes (e.g., $y = A+B+C+probability$). But according to neomechanics and its assumption of **infinity**, the equation should be $y=A+B+C+...\infty$. Nothing that happens in the Infinite Universe ever is the result of a finite number of causes. We are forced to lump the less significant causes into a factor we call "probability," but then it all comes down to a matter of interpretation: either probability is a singular cause, or it is not. By considering probability to be a singular cause, Copenhageners could keep the classical belief in finite causality along with popular assumption the universe was finite. Some of them subsequently went off the deep end—in no other part of regressive physics has solipsism become so entrenched. It got so bad that one often hears ridiculous claims like this: "A particle does not exist until I observe it."

The second violation, aether denial, fits right in with that claim. It also fits with the prevailing scientific world view: systems philosophy. In that vein, quantum mechanics attempts to explain the properties of tiny microcosms without proper attention to their surroundings. Quantum phenomena typically are attributed solely to the particles themselves. That is how particles strangely got to be seen as waves too. Of course, the phenomena explained by wave-particle duality are just the result of particles making waves in aether. That is how we get the "spookiness" of action-at-distance and other indeterministically interpreted phenomena sensationalized in the popular press. Here is an excellent 3-minute video of some overt propaganda supporting wave-particle duality.[246] The announcer, Morgan Freeman, follows the regressive script without fail. The shocking thing here is that Morgan, and everyone else, can clearly see the silicon particle floating on the liquid and making waves during its movement. He then declares that "they are particles *and* waves." In that vein, a ship and the waves it makes would have to be considered a "wave-particle."

I guess we could say: Duh?! Our ship at sea analogy does the same thing. Microcosms are microcosms, no matter how large or small they are. Like other things, quantum particles take up 3-dimensional space, influencing the macrocosm in which they exist. Morgan Freeman's conclusion is messed up. He must declare allegiance to wave-particle duality even though the

---

[246] http://go.glennborchardt.com/Freeman [Accessed 20170417.]

particles before his eyes are distinctly separate from the waves they produce. At least Morgan got one thing right: reality exists.

16.9 Why does matter prevent the transmission of aether waves?

We speculate that aether particles are the medium for light, the physical cause of gravitation and magnetism, as well as the precursor to baryonic matter. We also speculate aether particles are so tiny (3×10^{-19} cm) and universally ubiquitous they permeate everything. Gamma rays travel through the aether medium with wavelengths as short as 1.75×10^{-13} cm (the diameter of a single proton). This is about 0.000002 nm, while yellow light has a wavelength of about 570 nm.

Even though aether particles permeate all baryonic matter, the wave motion that travels through the aether medium eventually stops. For instance, light barely penetrates the ocean for more than 200 m. Most gamma radiation barely penetrates lead shielding that is about 40 cm thick. This is because the molecules within these media are microcosms that absorb and emit EM radiation per the $E=mc^2$ equation (Figure 24). The absorption is greatest when a microcosm first forms; the emission is greatest when a microcosm dissipates. As an aside, this is what happens when plants absorb sunlight. They may not emit that motion for a long time, perhaps not until they ignite in your fireplace. In general, the rate of absorption is a function of the density of the baryonic matter, with water being 1 g/cm^3 and lead being 11.34 g/cm^3.

16.10 Why would a finite particle be impossible?

I touched upon this in the "god particle" discussion and alluded to this many times before. As the atomists proved, the idea of a finite particle giving mass to all microcosms is only that—an idea. The required "perfectly solid matter" of the absolutist will never be found. One reason stemming from the above discussion is the fact the E=mc² equation must apply to all microcosms (Figure 24). If that equation should fail, then any such microcosm would lose its required connection to the macrocosm. It could not absorb or emit motion from or to its surroundings. Without those interactions, it could not evolve and would be as stagnant as the solid, inert atoms of the atomists. No such particle could exist in the Infinite Universe.

16.11 Does dark matter exist?

Yes. Dark matter appears to be real. For 80 years, astronomical observations have shown many galaxies behave as though they contain many times more mass than the amount attributed to the luminous matter seen with telescopes. As most of you know, gravitation between any two objects is dependent on their masses. Thus, if a small galaxy were to pass by a large galaxy, it would be pushed toward the larger galaxy per Newton's equation for gravitation. The curvature of its path would be dependent on its mass and the mass of the larger galaxy. Astronomers calculate the mass of a galaxy by estimating the number of stars it has. For instance, our own star, the Sun, has a mass of 2×10^{30} kg. There are about 400 billion stars in our galaxy, the Milky Way. Thus the visible mass is about 8×10^{41} kg. The total mass is about 3 times that. So, two thirds of the Milky Way is nonluminous. By studying rotation rates, Vera Rubin found some galaxies to have up to ten times as much mass as indicated by their luminous constituents.[247]

But what is this nonluminous dark matter? In our book, "Universal Cycle Theory: Neomechanics of the Hierarchically Infinite Universe," we speculated that dark matter is ordinary

---

[247] Rubin and Ford, 1970, Rotation of the Andromeda Nebula from a Spectroscopic Survey of Emission Regions; Rubin, 2000, One Hundred Years of Rotating Galaxies.

(baryonic) matter. Vortex theory implies that rotation produces particle size segregation following Stokes Law. As you can see in our demonstration video,[248] large, dense particles in a vortex are pushed toward the center first, with small, less dense particles following. This implied that there should be non-luminous stuff surrounding spinning galaxies and galactic clusters. This stuff could be planets, asteroids, rocks, molecules, atoms, or aether—anything that has mass. You would not be able to see any of these small, distant objects with a telescope. Small microcosms generally are magnitudes more plentiful than large ones, so numerous unseen exoplanets would contribute mightily to the mass of a rotating galaxy. These objects indeed would be "dark matter," but would they be sufficient to produce all the effects known for dark matter? No. For one thing, dark matter also is associated with cosmological objects that are not vortices—they are not rotating much. For instance, globular clusters are like elliptical galaxies—they are nearly spherical with little spin, but with masses up to three times indicated by their illumination.[249]

With respect to dark matter, one popular report claims we are getting close to discovering the nature of the "mysterious substance believed to hold the cosmos together."[250] This amendment to the Big Bang Theory had better be good, for it involves another $2 billion of your tax money. According to the authors, physicists declare they will soon know whether the missing matter "could be the strange and unknown dark matter or could be energy that originates from pulsars."

Well, it sure will not be energy, since energy does not exist. I find the silly comment about the "mysterious substance that is

---

[248] http://scientificphilosophy.com
[249] Rejkuba and others, 2008, Masses and M/L Ratios of Bright Globular Clusters.
[250] Heilprin and Borenstein, 2013, Scientists find possible hint of dark matter.

believed to hold the cosmos together" to be nevertheless intriguing. At first thought, the Infinite Universe does not need anything to "hold it together." Only the finite universe of the Big Bang Theory would need that. This is a subtle admission that a push, rather than a pull, would be necessary to do the "holding." Behind it is the sneaking suspicion the imaginary "gravitational attraction" might not be enough to stop universal expansion.

Our early work on the physical cause of gravitation concentrated on vortex formation. I have since come to realize that rotation is not required. I now speculate that dark matter is the decelerated aether that surrounds all baryonic matter. See the section on Aether Deceleration Theory.

16.12 Is matter lost during atomic fission?

No. Nonetheless, extensive propaganda around Einstein probably has most people thinking that matter or mass can be turned into energy. This is not surprising when the pillars of society keep pushing that shibboleth along with the required aether denial. A good example is this one from Forbes: "Einstein Was Right: You Can Turn Energy into Matter."[251]

All chemical reactions, including both fusion and fission, involve either the absorption or emission of motion (Figure 24). During absorption, the submicrocosms within a microcosm speedup, producing an increase in the momenta of the submicrocosms and an increase in resistance to acceleration—an increase in mass. Emission does just the opposite, with the emitted motion producing an acceleration of supermicrocosms (e.g., aether particles)—a decrease in mass. Per **conservation** the amount of matter (and motion) in the universe remains the same before and after these reactions.

---

[251] Rodgers, 2014, Einstein Was Right.

Chapter 17
Cosmology

In cosmology, Infinite Universe Theory closes many indeterministic doors and opens innumerable deterministic ones. For instance, simply giving up the corpuscular theory of light and its ad hoc assumptions severs the connection between physics and cosmogony. The chapters below recapitulate much of the previous discussion. The arguments made here follow from the fundamentals already laid down. In particular, they follow one of the main themes of Infinite Universe Theory: No paradoxes or contradictions are allowed.

17.1 Is the universe expanding?

No. The Infinite Universe can do much, but it cannot expand, because the Infinite Universe is already full. The required empty space is only an idea, an idea that perfectly empty space is possible. Again, the Infinite Universe can do many things, but the production of nonexistence (empty space) is not one of them. The expansion hypothesis depends on two important indeterministic assumptions: 1) the cosmological redshift is a result of the Doppler Effect and 2) our apparent location at the center of the supposed expansion is a space-time effect.

As shown previously, the Doppler Effect occurs when an observer moves toward or away from a propagating wave. As in the boat example, the Doppler Effect only occurs when there is a wave within a medium. That is why everyday explanations of it always involve a medium and why aether deniers need to gloss over the medium problem when discussing light. That is why, in wave-particle theory, photons are alleged to bring their own waves with them as they travel through supposedly empty space.

Why waves form

Waves are produced when the motions of a particular microcosm are transmitted to the macrocosm—imperfectly, as they must be. If the idealist's solid matter existed, then this motion would be transferred instantaneously—there would be no waves. Nonetheless, we get close to this kind of motion transfer when we poke a cue stick at a billiard ball. The motion of our hands is transferred to the stick and then to the billiard ball as if instantaneous even though we know there must be atom-to-atom contact within the stick for that to occur. Motion is always transferred microcosm-to-microcosm. This always involves some delay, even if only for a fraction of a nanosecond. There is wave motion within the stick, although the rigidity of the stick may prevent us from noticing it. For a medium with less rigidity, such as water, for instance, wave motion becomes obvious. Above all, though, the velocity of the wave motion and its character is dependent on the properties of the medium. You can hit the water as hard as you wish, but the waves will still take their own sweet time; you can turn up the bass as loud as you wish, but sound waves in air will still travel at only 343 m/s.

Wave components and 3D:

Having a great deal of freedom, the molecules in media such as water and air cannot be forced to transmit all their motion unidirectionally. Only an impossibly direct hit between microcosms could do that. Nonetheless, when the transfer is nearly direct, an L (longitudinal) wave is produced in the direction of motion (also called a pressure wave or P wave). In the case of light, this is called *radiation pressure*.[252] When the transfer is more oblique in extra squishy media, the microcosms squirt off to the side. This produces a T (transverse) wave, which

---

[252] http://go.glennborchardt.com/radpres

on average appears perpendicular to the direction of motion. This is often called a shear wave, because the microcosms, rushing past each other, tend to be impeded on the sides touching each other. For each pair, the resulting drag forces some of the motion to be absorbed as spin when one of the microcosms ends up rotating clockwise and the other ends up rotating counter-clockwise.[253] Because the universe is three dimensional, the motions produced by waves display three components: 1) back and forth, 2) up and down, and 3) side to side. Good thing light is transverse wave motion; otherwise, polarized glasses would not work.

The ability to transmit either L or T waves depends on the characteristics of the medium. Air transmits mostly L waves—that is why the diaphragm on your base speaker goes back and forth. Water's surface transmits mostly T waves—that is why the water's surface goes up and down, with little horizontal motion. Metal transmits both L and T waves. Light is transmitted mostly as T waves; at first glance behaving as a solid or liquid instead of the quasi-gas we assume aether to be. My own speculation on this considers the probable shape of the aether microcosm (see Figure 40). The molecules in gases, such as the N_2 and O_2 in air, are spherical, allowing mostly direct collisions between molecules, resulting in L waves. Aether particles therefore must not be spherical. They probably are disc shaped like the other vortices making up our hierarchically Infinite Universe.[254] Discs would rarely collide directly—most collisions would be perpendicular to the direction of motion, as we see in T waves.

The idea the universe is expanding is entirely dependent on Einstein's Untired Light Theory and the regressive interpretation

---

[253] http://go.glennborchardt.com/wavemotion
[254] Puetz and Borchardt, 2011, Universal Cycle Theory.

of the cosmological redshift. Big Bang Theory could not exist without it.

Redshift Primer

There are four types of redshift and a fifth that is bogus:

Type I Redshifts Produced by the Motion of the Observer

With the advent of relativity, there has been much confusion about the Doppler Effect so the following bears repeating. The Doppler Effect is simple, involving two points and the wave motion between them. Type I redshifts are the easiest to explain because you can produce them on any quiet body of water. If you move toward a wave-making machine (e.g., boat in action), you will encounter the waves it produces much more quickly. This is an observer-produced blueshift (waves appear sooner than normal). If you did not account for your own motion, you might say the wavelength of the waves was magically less than normal.

If you move away from the source of the waves, you will encounter the waves it produces much more slowly. This is an observer-produced redshift. The redshift observer will see the waves as further apart (an increase in wavelength) and will count fewer of them per minute (a decrease in frequency) than will the folks on shore.

Type II Redshifts Produced by the Motion of the Source

If the source moves toward you, it produces waves closer together than normal, as seen by the object moving left in Figure 52. This is a source-produced blueshift. Unlike the observer-produced blueshift, these waves will be closer together than normal because the motion of the source closes part of the usual gap between the waves. Both you and I will agree the waves become closer together as the source moves toward us. A redshift is produced when the source moves away from us. In this regard, light from Andromeda is blueshifted, because it is moving toward us faster than redshifts produced in other ways.

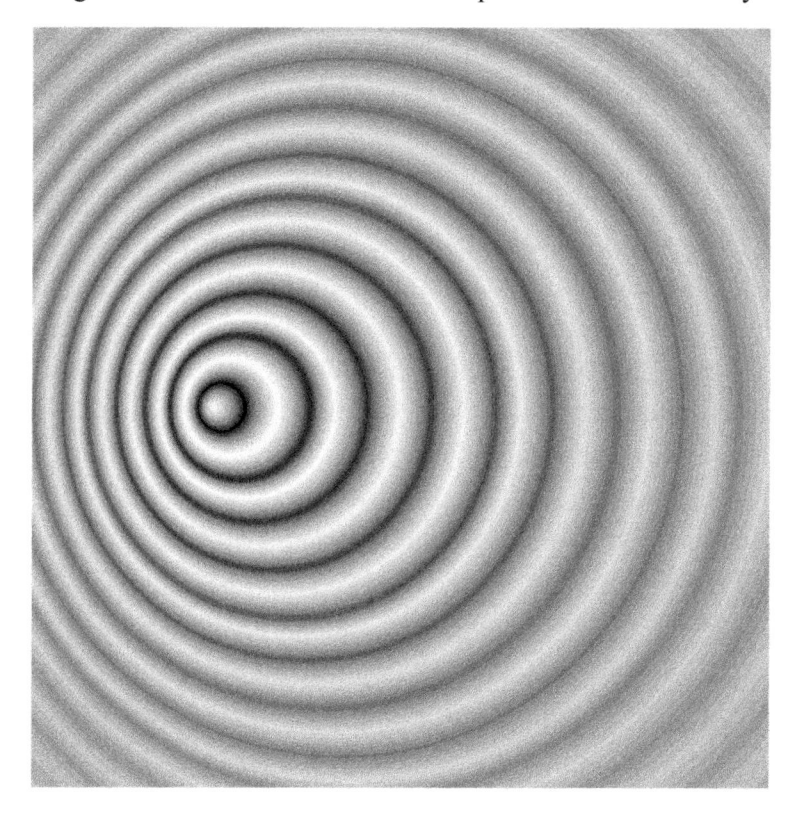

Figure 52 When the source is in motion to the left, wavelengths are blueshifted (shortened) to the left and redshifted (lengthened) to the right.[255]

Type III Redshifts Produced when Light Leaves a Cosmic Body

These are otherwise known as gravitational redshifts, which were discussed in Chapter 15.5

---

[255] http://go.glennborchardt.com/doppler

Type IV Redshifts Produced when Light Travels Great Distances Through the Aether

These are otherwise known as cosmological redshifts, which are discussed below in Chapter 17.3.

Type V Redshifts Thought to be Produced when Space Expands

Type V redshifts seem to be an ad hoc designed to save the Big Bang Theory from the general rule that redshifts can only occur in a medium. According to Einstein, light is a special particle that travels through perfectly empty space, carrying its waves along with it. Cosmogonists finally realized they needed to add something to the theory to maintain the fiction that the universe was expanding. That something was the absurd idea space itself was expanding. How and why something that was supposedly perfectly empty could expand is not clear.

For expansion to occur, at least two things must diverge from one another, so space cannot be perfectly empty. For instance, a balloon not filled with air does not expand when it is moved to high altitude. A balloon filled with air expands at high altitude because the nitrogen molecules within move apart as their collisions find less resistance from the outside air pressure.

This particular ad hoc brings its own contradictions with it. The analysis goes like this:

Space needs to be perfectly empty. Otherwise, Einstein's bogus Untired Light Theory, which is responsible for the expansion hypothesis would not work.

Cosmogonists would maintain it is not space that is expanding, but space-time.

Space-time has a material character because General Relativity Theory considers time to be a dimension.

This imagined material character of space-time is what allows it to expand in the same way aether deniers imagined it could compress and decompress in the LIGO experiment.

This particular bogus ad hoc is no worse than the idea the whole universe could explode out of nothing.

17.2 Can the Doppler Effect occur without a medium?

No.

17.3 What causes the cosmological redshift?

The cosmological redshift (Figure 11 and Figure 53 and supposedly indicates the universe is expanding. Once we confirm the true cause of this type of redshift, it will mean the end of the expansion hypothesis and the Big Bang Theory. According to neomechanics, we assume the cosmological redshift cannot be a result of any one of the other types of redshift. Of course, in an Infinite Universe filled with an aether medium, having microcosms moving in all directions, there still would be Doppler Effects. Half would be blueshifts and half would be redshifts, because each of those cosmic bodies either would be coming toward us or going away from us. Now, the cosmological redshift cancels out some of the blueshift of light from objects moving toward us, and adds to the redshift of light from objects moving away from us. The cosmological redshift also must include the small "gravitational" redshift produced by all luminous objects as their light encounters increased aether pressure when it leaves the necessarily baryonic-rich environs of the source.

The main contribution to the cosmological redshift obviously reflects the great distances involved. In neomechanics, we realize no microcosm or motion of microcosms could travel from point A to B without losses. Perfect transmission of matter or the motion of matter, like perfectly empty space, is only an idealist's dream—a dream Hubble himself refused to accept. He ultimately rejected the oft-repeated claim by regressive physicists and cosmogonists that he discovered the universe was expanding. This put him squarely in the "tired light" camp, which eschews the perfect light transmission of the indeterminists. Regressive physicists commonly malign the tired light theory even though it is an everyday occurrence on Earth. For instance, it is impossible to transfer electricity across the country without losses. The hum you hear when you are near a high voltage transmission line represents some of that loss. Only if the wire consisted of the idealist's "perfectly empty space," would it be capable of transmitting electro-magnetic motion without losses.

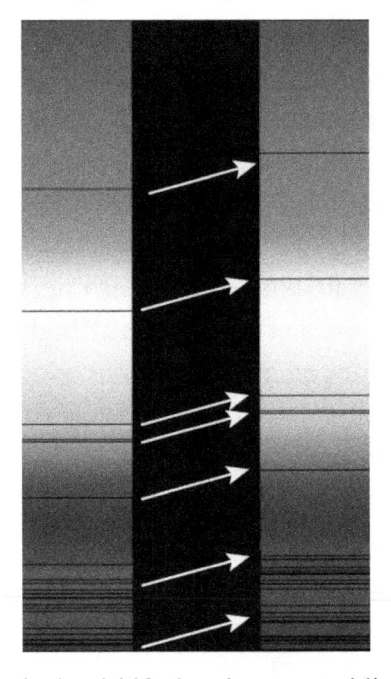

Figure 53 Cosmological redshift showing spectral lines for various elements being shifted to the red, long wavelength, low energy end of the spectrum. Credit: Georg Wiora.[256]

---

[256] http://go.glennborchardt.com/Wikiredshift. Georg Wiora (Dr. Schorsch) created this image from the original JPG. Derivative work:Kes47

Now, in seeking the cause of the cosmological redshift, we need not concern ourselves with Einstein's corpuscular theory. That is because we believe Sagnac showed light is motion: a wave in the aether. For light to be transmitted as a wave for over 13 billion years without losing motion, each of those waves must be reproduced perfectly. That simply cannot happen. Any delay whatsoever, even if fractions of a nanosecond would result in longer wavelengths. There is no mechanism by which a wave could become shorter and gain energy, but the many ways it could become longer and lose energy. Waves are always made up of septillions of individual microcosms that transfer the motion from microcosm-to-microcosm as seen in Wikipedia demonstrations.[257] Most importantly, each of those collisions is susceptible to the six interactions I outlined in the neomechanics chapter. Because aether particles are extremely dense (Table 11), the absorption of motion internally would be slight and would not be obvious over short distances. Note there are claims sound waves are redshifted over distance even though the effect is tiny and seldom noticed.[258] This is in tune with neomechanics, in which all Second-Law collisions between microcosms in all media must result in the absorption of at least tiny amounts of motion in addition to the usual particle-to-particle acceleration. The result must be a reduction in energy, which we see as an increase in wavelength over distance. Of course, this effect is superimposed upon redshifts and blueshifts produce by the Doppler Effect and any other effects that change wavelength. In

---

(File:Redshift.png) [CC BY-SA 2.5 (https://creativecommons.org/licenses/by-sa/2.5), GFDL (http://www.gnu.org/copyleft/fdl.html) or CC-BY-SA-3.0 (http://creativecommons.org/licenses/by-sa/3.0/)], via Wikimedia Commons.

[257] http://go.glennborchardt.com/wavemotion
[258] http://go.glennborchardt.com/RSsound

any case, the only thing preventing us from understanding the cosmological redshift is aether denial.

In "Universal Cycle Theory," we speculated that aether pressure is highest in intergalactic space. As noted in the discussion of the "gravitational redshift," this high pressure allows for faster than c transmission, which would produce at least some of the cosmological redshift. The redshift formerly attributed to the Doppler Effect or to "expanding space" is a function of the total distance light has traveled through intergalactic space. That is why the cosmological redshift increases in all directions, giving us the impression we are at the center of the universe. But, as with all media, we do not expect the properties of aether to be the same everywhere even though it appears ubiquitous. In "Universal Cycle Theory," we emphasized that there are no true constants in nature; just as there are no identities per our assumption of **relativism.**

17.4 Are there galaxies more than 13.8 billion years old?

Yes. First, the actual ages of the oldest galaxies are by no means certain. While redshift equivalents point to 13.8 Ga (billion years old), still other analyses involving the Cosmic Microwave Background claim the universe is 84.5 billion light years in diameter.[259] Do not ask me how that can be—it is not my theory. I guess it is supposed to involve space itself expanding at superluminal velocity—another ad hoc to keep the Big Bang Theory alive. As mentioned early on, without expansion, the 13-Ga galaxies at observation limits would now be over 26 Ga. The Milky Way itself is only 13.7 Ga. In our book, "Universal Cycle Theory," we estimate it will take another

---

[259] Vaudrevange and others, 2012, Constraints on the topology of the Universe.

37,000 trillion years for the Milky Way to mature.[260] With humanity's tendency to be homocentric and with our location being in a relatively young galaxy, we are accustomed to young ages.

Second, the assumptions of the Big Bang Theory demand all cosmological ages fit within the 13.8-Ga interval. That is a special problem when we realize the Milky Way galaxy is part of the Local˙ Group of galaxies, which is part of the Virgo Supercluster. This hierarchical nature seems to imply much older ages than calculated by Big Bang theorists. The youthful Milky Way must have had considerable mass before gravitation could push it toward the other galaxies in the Local Group. Similarly, the youthful Local Group must have had considerable mass before gravitation could push it toward the other galaxies in the Virgo Supercluster.

Third, we do not expect to find infinitely old microcosms in the Infinite Universe in any case. Each microcosm, no matter how large or small—including the observed universe—must have a finite age. That is described by **complementarity**. In the Infinite Universe, things come into being from other things and go out of being into other things. Microcosms have lifetimes from nanoseconds to trillions of years, but theoretically, they all have lifetimes. The Higgs Boson apparently formed in the CERN accelerator, had a lifetime of 10^{-22} seconds, while the proton has a half-life greater than 1.29×10^{34} years."[261]

Fourth, no single type of microcosm is regenerated endlessly. That is because everything in the Infinite Universe is constantly moving, including the macrocosm contributing to the formation of any particular type of microcosm. With **relativism**, we assume

---

[260] Puetz and Borchardt, 2011, Universal Cycle Theory, p. 172.
[261] http://go.glennborchardt.com/protondecay

there can be no identical microcosms. No two galaxies, like no two snowflakes, are identical.

Fifth, according to our assumption of *infinity*, the universe is infinite, both in the microcosmic and macrocosmic directions. This implies there is no smallest microcosm and there is no largest microcosm. As we explained in "Universal Cycle Theory," a microcosm in the hierarchically Infinite Universe is always part of a still larger microcosm—as demonstrated by galaxy clusters, super clusters, the Huge Large Quasar Group (up to about 4 billion light years across), and the Local Mega-Vortex Stephen and I speculated about in our book (Figure 54). Picking on any one of these or any other favorite microcosm for regenerating the universe cannot possibly work. Such an attempt at perfect regeneration would be to assume *finity* instead of *infinity*. That is what the atomists did in assuming atoms were identical spherical particles filled with perfectly solid matter. As we showed in our book, there is no reason for the hierarchy to start or stop at any particular part of the scale from microcosmic infinity to macrocosmic infinity.

Figure 54 Cover of our book showing the observable universe rotating around the "Local Mega-Vortex."[262] We based on observations that galaxy clusters flow in a preferred direction, as if they were revolving around some sort of massive celestial body outside the observable universe.[263]

---

[262] Puetz and Borchardt, 2011, Universal Cycle Theory. Neomechanics of the Hierarchically Infinite Universe.
[263] Kashlinsky and others, 2010, A New Measurement of the Bulk Flow of X-Ray Luminous Clusters of Galaxies.

17.5 What causes the Shapiro Effect?

As the holy grail of efforts to undermine the Big Bang Theory, there have been numerous attempts to pin the cosmological redshift on tired light, without much success. One example is the Shapiro Effect, which is a time delay observed for light as it passes through the atmosphere of a cosmic body.[264] The Shapiro Effect, however, is simply another manifestation of the slowing of light velocity due to refraction and to the misnamed gravitational redshift. As I explained above, light slows down and is blueshifted as it nears a massive body and enters the atmosphere. It speeds up again and is redshifted when it leaves the massive body. For light coming toward us and passing another cosmic body, the blueshift and redshift cancel each other out, although the atmospheric entrance and exit takes extra time, causing the "Shapiro Delay." It is true light probably will traverse many of these atmospheres during its 13.8-Ga travels.[265] Again, that would cause a time delay, but would not cause a redshift. That is because the seconds lost by light via refraction upon entry to an atmosphere would not be completely regained upon exit. The same phenomenon occurs when light enters and leaves a pool of water. While in water, light travels at 225,000,000 m/s, but at 300,000,000 m/s when it leaves. If the atmosphere of a planet was pure water, the resulting "Shapiro Delay" would be great. If there was no atmosphere, the delay would be closer to zero, baring the effects of aether deceleration, of course.

---

[264] Shapiro and others, 1985, The Search for Gravitational Waves; Thacker, 2013, The Shapiro Effect.
[265] Thacker, ibid.

17.6 Will the universe suffer "heat death"?

No. This is a logical offshoot of a misinterpretation of the Second Law of Thermodynamics (SLT) by systems philosophers who invariably assume *finity*. The Sixth Assumption of Science, **complementarity** resolves the SLT-order paradox as I pointed out as early as 1984.[266] The indeterministic assumption is just the opposite: *Noncomplementarity*. The Second Law of Thermodynamics states that all isolated systems eventually run down. In the regressive interpretation, constituent matter supposedly is converted into "energy," which escapes the isolated system as unusable heat. Another way of stating it from a mechanical viewpoint is that the constituents of the system will diverge or expand into its surroundings via their own momenta. Either way, both interpretations fit the expanding universe of the Big Bang Theory with divergence being assumed greater than convergence.

But, if you have been following my argument in favor of univironmental determinism, you know that no systems are ideally isolated. If they were, then the Second Law of Thermodynamics would not even work. The system's boundaries would have to be "leaky" or stretchable for the heat or matter to escape its confines. Of course, if one treats the universe as a finite, isolated 3-D system, one could argue it would expand into the perfectly empty surroundings, that, for some reason, escaped the imagined creator's touch. On the other hand, if one treats the universe as a finite, self-contained 4-D system with no surroundings as the Big Bangers do, one could imagine its expansion without having to imagine its surroundings being

---

[266] Borchardt, 1984, The scientific worldview. [Early manuscript version of the 2007 book. Also, an early version of the resolution was rejected by *Science* in 1980 and finally published as Borchardt, 2008, Resolution of the SLT-order paradox.]

empty. In either case, one must use the indeterministic assumptions of *finity* and *noncomplementarity* and the deduction divergence could occur without an equal amount of convergence.

There is a bit of truth to the correlation of expansion with death. Except for the Infinite Universe, all microcosms come into being via convergence and undergo death via divergence. The assumptions the universe had a beginning and will have an ending are logically derived from our everyday observations of everything in the universe. The only problem is that they cannot apply to the to the universe as a whole.

I guess the heat death hysteria may be fading away as the standard Big Bang Theory comes under attack and modifications are suggested to handle some of its major contradictions. Cosmogonists are moving slowly toward Infinite Universe Theory by suggesting oxymoronic solutions called "parallel universes" or "multiverses," while holding fast to the indeterministic assumption of *finity*.

Each of those oxymoronic "universes" is based on the expansion hypothesis, which, in turn, is based on Einstein's Untired Light Theory. These steps out of the cosmogonic box are admirable and perhaps one of them could be the "super great attractor" responsible for the galactic flow discovered by Kashlinsky and others.[267] It is true that, in the future, Infinite Universe Theory always will be subject to change. For instance, Stephen Puetz and I presented a hierarchical version in "Universal Cycle Theory" in which the observed universe revolves around a "Local Mega Vortex." That is highly speculative, but we consider it superior to the oxymoronic alternatives. Nonetheless, we stand by the view the universe is eternal and extends infinitely in all directions.

---

[267] Kashlinsky and others, 2010, ibid.

17.7 Why is a finite universe impossible?

As seen in previous chapters, there is much logical support for the belief that the universe is infinite. On the other hand, there are many reasons indicating that a finite universe is impossible. I will relist some of them here along with some of the ad hocs you will need if you still wish to continue your belief in cosmogony.

To believe that the universe had a beginning, one must also believe in nothing, perfectly empty space. But as I mentioned many times, perfectly empty space is only an idea. It is one end member of the matter-space continuum. As with all idealizations, perfectly empty space cannot exist.

Baring Einstein's "time is a dimension" ad hoc, a finite universe would have to be surrounded by perfectly empty space, which cannot exist because it is an idealization.

You must reject univironmental determinism, the universal mechanism of evolution in which what happens to a portion of the universe is determined by the infinite matter in motion within and without. Contrary to current systems philosophy, each microcosm must have a macrocosm filled with matter in motion.

Gravitation would be impossible in a finite universe. This is obvious from the failure of systems philosophers to discover the physical cause of that process. For over three centuries, myopic believers in a finite universe have promulgated the magical, acausal "attraction" theory, not appreciating the fact that gravitation is univironmental. As shown by Aether Deceleration Theory, the very shape of a cosmic body is dependent on causal pushes (F=ma) from its surroundings. Without aether pressure, each microcosm would explode into the macrocosm per the Second Law of Thermodynamics.

To accept the assumption of *finity*, you must reject the Sixth Assumption of Science, **complementarity** (All things are subject to divergence and convergence from other things). That is because a finite universe would be subject to the Second Law of Thermodynamics without its complement. The Second Law of Thermodynamics is a law describing divergence, essentially claiming that isolated things only can come apart as they undergo increasing entropy. It predicts that the supposed finite universe must undergo eventual "heat death." Nonetheless, all around us we see birth as well as death—there are just as many things coming together as well as coming apart. The resulting "SLT-Order Paradox" cannot be resolved unless we assume the universe is infinite.[268]

A finite universe can have no physical cause. That is because physical causes are defined by Newton's Second Law of Motion as the result of collisions described by the F=ma equation. On the contrary, each portion of the Infinite Universe is subject to mechanical collisions from elsewhere. An ad hoc requiring a supernatural being capable of creating over two trillion galaxies is unnecessary.

The currently popular cosmogony hypothesizes a grand explosion, a supreme divergence that resulted in the creation of everything in existence. This is particularly incongruous in view of the fact that the creation of anything we know of is the result of convergence, not divergence. Only in the minds of regressive physicists and cosmogonists can things be brought together by blowing them up.

---

[268] Borchardt, 2008, Resolution of the SLT-order paradox.

PART IV: CONCLUSIONS

Chapter 18

Predictions, Persistence, and Requiem

18.1 Predictions of the New Theory

The Big Bang Theory will be replaced by the Infinite Universe Theory.

Neomechanics will replace classical mechanics and the indeterministic interpretations of relativity.

Matter-motion terms such as momentum, force, energy, and space-time will be recognized as calculations rather than as matter or motion.

The universe is not expanding.

Light is a wave in the aether.

Aether particles have disc shapes responsible for transverse motion in the aether medium.

Light velocity increases with distance from baryonic matter.

Gravitation is the result of aether pressure, which decreases with nearness to baryonic matter due to aether particle deceleration.

The gravitational redshift, now considered proof of relativity, is a result of increasing light velocity due to increases in distal aether pressure.

The cosmological redshift is the result of the microcosmic absorption of motion, which is what happens to all wave motion over distance.

Time will be recognized as motion and thus incapable of dilation or dimensionality.

Theories touting more than three dimensions will be regarded as quaint and unrealistic.

Improvements in instrumentation soon will result in the discovery of cosmological objects older than 13.8 billion years.

18.2 Paradigmatic Persistence of the Big Bang Theory

We have been hearing about the Big Bang Theory for more than half a century. As we have seen, some of the conclusions of that theory and its assumption of *finity* are fantastic. You might like that, just as you might like science fiction or some promised, shiny abode in the sky. After all, believers want to believe. There is a certain feeling of comfort and well-being when one is ensconced in a paradigm that brings along with it *certainty* and financial reward. This section examines the reasons for the persistence despite the multiple contradictions of the current cosmogony.

The philosophical struggles over its dubious founding assumptions have long ago subsided with the Big Bang Theory having been declared the victor. Pointing out all those unrelenting contradictions once again, as I do in this book, is unlikely to convince true believers. The neurological pathways in our brains seem to form a sort of railroad track that becomes ever more embedded the more we use that track. This is why it is increasingly difficult to get "sidetracked" or "think outside the box" as we age. It is why it is so difficult to give up old habits and to establish new ones. It is why companies trying to be innovative hire young workers. It is why those who have believed for many decades that the universe exploded out of nothing are likely to continue along that track. It is why those who have repeated conventional mantras such as "there is no aether," "there is no aether," "there is no aether" 10,000 times are not likely to utter or believe that its opposite could be true despite Einstein's recantation.

It is also why those who never believed in the Big Bang Theory in the first place are unlikely to start believing in it no matter how long they live. The critical juncture for these two tracks occurs at an early age, when one is considered to have an "open mind." The selection is determined univironmentally, that is, by the microcosm of the individuals and the macrocosm in which they exist. Because no "open minds" really can exist, each of us is predisposed toward certain decisions based on the presuppositions we learned still earlier in life. Continual repetition of cosmogonic views, whether it is via the Big Bang Theory, Genesis, or New Scientist sends one down the same indeterministic road. The macrocosm contains peer pressure and financial pressure that helps to guide us along one track or another.

Although it may seem that way, belief in the Big Bang Theory, like the belief in the 72 virgins, has little to do with bamboozlement. It has everything to do with the reiteration of fundamental presuppositions of which the tracked individual is unaware. That is why "The Ten Assumptions of Science" are so revealing. Even though there are numerous paradoxes and contradictions along the track, the belief in the Big Bang Theory is purely logical. In addition to subconsciously favoring the indeterministic opposites of "The Ten Assumptions of Science," these derivative assumptions must be ingrained within believers:

1. Aether does not exist.

2. Space is perfectly empty.

3. There is a finite particle.

4. The universe is finite.

5. The universe had a beginning.

6. Time is a dimension.

7. There are four dimensions.

8. Space-time exists.

9. Curved space exists.

10. Gravity is a pull.

11. Fields do not contain matter.

12. Light is a particle.

13. Photons contain waves.

14. The universe is expanding.

15. Forces exist.

16. Energy exists.

17. And many, many more…

Within regressive physics, sidetracks cannot use more than tiny modifications of these derivative assumptions. None must contradict Einstein, who is always right. But, as I have maintained throughout our books, each of these derivative assumptions is false. The Big Bang Theory is the product of a long-standing logical evolution among honest folks who grew up believing that contradictions are a necessary ingredient in philosophy. To true believers, the Big Bang Theory paradigm is not absurd and certainly not some kind of conspiracy or bamboozlement. It is an outgrowth of what they already accept as truth. Replacing the Big Bang Theory with the Infinite Universe Theory will not be easy. We can do the math, but we will not get on the right track without first starting with the correct assumptions.

Obviously, despite its rejection of commonsense and the presence of numerous contradictions, Einstein's relativity and the Big Bang Theory have achieved tremendous popularity. In "The Structure of Scientific Revolutions" Thomas Kuhn used the word "paradigm" for any theory or set of related theories used to advance a particular line of thought during a particular historical period.[269] Each paradigm generally is the victor in a long hard struggle to replace old ways of thought made outdated by accumulating data and increasing exposure to the Infinite Universe. Each paradigm has firmly held underlying assumptions that tend to fade from view with each success of the paradigm. A paradigm provides answers and predictions that satisfy its participants well enough that they can continue in the discipline. Despite ever-present claims to the contrary, every paradigm is economically dependent—its practitioners cannot live on air. In addition, each paradigm needs political and philosophical support. The Big Bang Theory has that in spades.

Paradigms are efficient, providing closure for topics no longer discussed. Chemists, for instance, do not wake up trying to decide whether to include the possibility of miracles in the day's analyses. Geologists studying reverse faults (where old rocks are pushed over young rocks) do not need to choose between plate tectonics and the expanding earth hypothesis. Rocket scientists do not need to review the literature on the flat-earth, geocentric, and heliocentric theories. Think of all the time wasted if we always had to review first principles before doing anything!

Of course, as new data accumulate, the need for modification becomes increasingly apparent.[270] Most of these modifications, however, are part of what Kuhn calls "normal" science. The modifications are performed by salaried practitioners familiar with the expectations of the paradigm. As with today's internal combustion engine, an elderly paradigm tends to accumulate all kinds of add-ons that keep it running smoothly in the face of changes in the environment. Data that do not fit the paradigm remain unpublished or ignored. This is only human. It is what we

---

[269] Kuhn, 1962, The structure of scientific revolutions.
[270] That is why this book will require frequent revisions.

do in every debate—emphasize points that support our view and deemphasize points that do not.

Kuhn considered the change from one paradigm to another as "revolutionary science." Among his most startling observations: such drastic changes are seldom performed by the practitioners. They may recognize the need for it, but will be unable to meet the requirements. Even Einstein, a patent officer, was initially an outsider to mainstream physics. Recently there has been halting movement away from the finite model, with reformers inventing theories involving "multiverses" and "parallel universes" (

Figure 55). Obviously, these terms are oxymoronic, just like the term "island universe" was before galaxies were confirmed as portions of the one and only universe. It is good to see mainstream folks getting out of the box, even if they remain unschooled in the language. Maybe such inventions solve some problems, but it is clear that the promulgators have little clue as to what really is required. To be true to relativity, fields still need to be immaterial and to be true to Big Bang Theory, every "universe" still must explode out of nothing, whether parallel or not. When it comes to producing the next paradigm shift, professional physicists and cosmogonists need not apply.

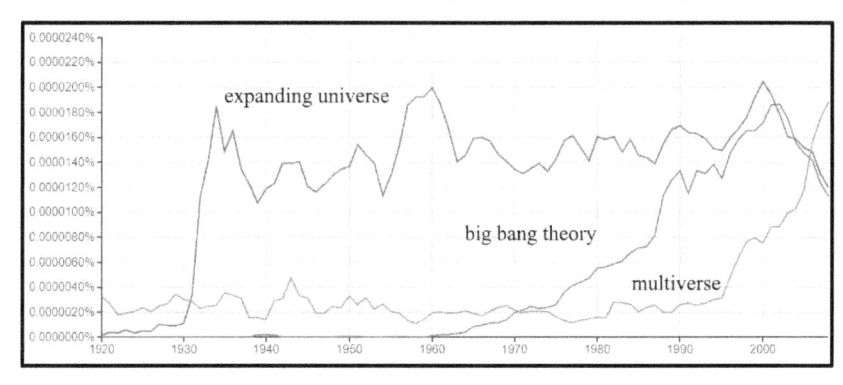

Figure 55 Google Ngram for books that use the phrases or words: expanding universe, big bang theory, or multiverse. Note that the

popularity of the expanding universe interpretation preceded the Big Bang Theory by over three decades.[271]

So if practitioners are unqualified to produce the next major paradigm shift, then who is? This one is especially tough because it has a built in contradiction: One needs to understand physics and cosmology well enough, but not so well that one's life depends on it. As Upton Sinclair famously said: *"It is difficult to get a man to understand something when his salary depends upon his not understanding it."*[272]

Then too, practitioners within a paradigm invariably reject dissident views—that is their job. After all, atheists do not get to preach in church. This is not a conspiracy, as some foolishly claim. Each discipline, like each religion, must uphold certain norms. For instance, one could not be an evolutionary biologist without believing in evolution; one could not be a chemist without believing in chemicals. The peer review system protects these norms, rejecting manuscripts that step out of line. As Kuhn observed, thinking outside the box is encouraged, but stepping completely outside the box is not.

Both Kuhn and Collingwood[273] provided a few hints. It has nothing to do with some math tweaks that finally will make everything better. No, it has to do with the foundations of the paradigm, the fundamental assumptions responsible for the paradoxes and contradictions that may be evident to bystanders but not to practitioners. According to Collingwood, our fundamental assumptions exist as unconscious presuppositions until dragged into the light of day. Discovering these and

---

[271] Michel and others, 2011, Quantitative Analysis of Culture Using Millions of Digitized Books. See: https://go.glennborchardt.com/ngramBBT (accessed on 20181018).

[272] Sinclair, 1935, I, candidate for governor and how I got licked.

[273] Kuhn, 1962, The structure of scientific revolutions; Collingwood, 1940, An essay on metaphysics.

dragging them out, by definition, is not an activity practitioners are interested in pursuing. How many scientists want to learn that their careers have been spent barking up the wrong tree? Do you think that any of the cosmogonists would want to believe that?

Nonetheless, a new assumptive foundation is exactly what we need to replace the current paradigm. I feel confident that "The Ten Assumptions of Science" are a welcome start to developing that foundation. If you examine these assumptions and their opposites closely, you will see why we face such strong paradigmatic persistence today. Nearly everyone has been exposed to many hours of training based on the indeterministic opposites. Most still believe in free will. Most still believe that their salvation depends on a soul. Who doesn't believe that gravitation is a pull? Who doesn't believe that light is a photon? Most believe that the universe had a beginning. The propaganda that gets us to believe this stuff is all-pervasive. It is politically, economically, and philosophically driven. Sorry dissenters, but that will not change simply because someone discovered a new equation or found a math error in an old one.

18.3 Requiem for the Big Bang Theory

The paradigmatic shift needed to displace the Big Bang Theory will only come about when the greater society is ready for it. Like all change, our view of the universe depends greatly on how we view ourselves, and *vice versa*. Other scientific revolutions occurred in relatively cloistered communities, slowly spreading throughout the educated as communication allowed. This one will be different. It will involve the entire globe with communication at the speed of light. Still, the old, indeterministic assumptions survived for millennia. It will take a lot to convince superstitious folks that the universe has only two fundamental phenomena: matter and the motion of matter. After all, who wants to hear that this means that there will be no living after dying?

As always, great philosophical changes occur only in response to great changes in the material conditions of humanity. In this, the continuing Global Demographic Transition portends what is to come (Figure 56). Humanity's juvenile development occurred about the mostly unheralded Global Inflection Point, which we observed in 1989. That year the Earth reached its maximum fertility, adding 88 million people and reaching half way to its carrying capacity of 10 billion.

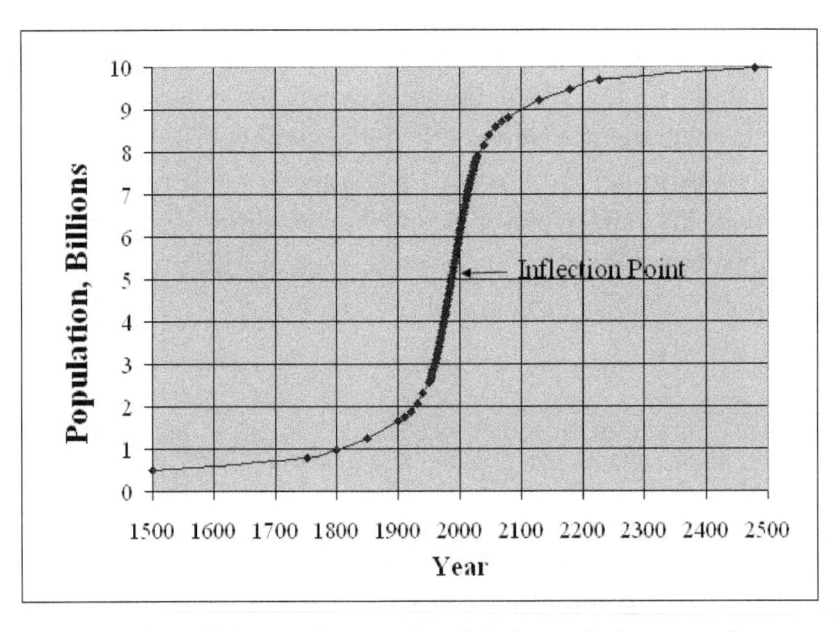

Figure 56 Sigmoidal growth curve for global population assuming perfect symmetry about the 1989 Inflection Point.

Although global population continues to expand, the rate of the expansion has declined to 77 million per year. Not coincidentally, the population explosion between 1950 and 2050 has been, and will be associated with the greatest global economic expansion we will ever undergo. It parallels the Industrial Revolution, the urbanization of the planet, and the rapid expansion of capital. This will continue, but instead of ever-increasing profits, we must contend with ever-decreasing profits as globalization turns Earth into an island. Instead of assuming mathematics to be the final arbiter of our understanding of the universe, we will assume the universe to be the final arbiter of mathematics. Instead of ever increasing economic growth and resource exploitation, the watchwords will be ever increasing economic stability and sustainability. Due to mechanization and robotification, there will be growth pains aplenty, but there will be no turning back. No high-tech panacea or interplanetary migration can circumvent the inevitable socialization of the entire planet.

This mirror image in Figure 56 was developed from historical estimates and 1950-1989 data from the U.S. Census Bureau.[274] Note that the curve correctly predicted today's 7.4 billion population. The figure shows that the global population will be about 8.5 billion in 2050. It will not reach the United Nations "medium projection" of 9.2 billion and certainly not their "high projection" of 10.8 billion by that time. It is unlikely to decline for centuries. What we are seeing in Figure 56 is nothing less than a reflection of the Global Industrial Revolution itself. The slowing population growth rate has been and will continue to coincide with the slowing economic growth rate.[275] By 2050, stress on the current economic system, geared to perpetual growth, will be intense, forcing global economies toward a more sustainable system. Because the effect on people's lives will be so dramatic, the switch from rapid growth to near-

---

[274] Borchardt, 2007, "The Scientific Worldview," p. 290.
[275] Piketty, 2014, Capital in the twenty-first century. [For the opposing point of view, see Magness and Murphy, 2014, Challenging the Empirical Contribution of Thomas Piketty's Capital in the 21st Century.]

zero growth will be accompanied by an intensification of the philosophical struggles between socialism and capitalism, determinism and indeterminism, and science and religion. These ferments within the greater society will call everything, including the Big Bang Theory into question. Since we already had our counterrevolution in physics and cosmology, there is nowhere else to go. The next revolution will be progressive. The demise of cosmogony will be slow in coming, but its demise will be an inevitable part of humanity's maturation. Once we flip from the indeterministic assumption of *finity* to the deterministic assumption of *infinity,* there will be no turning back. Unlike other scientific revolutions, we are not dealing here with a small portion of the universe, but all of it. Sure, there will be inevitable modifications to the theory, but the basic skeleton that sees the universe as infinite and eternal will remain. Welcome to the "Last Cosmological Revolution."

References

Abu-Bakr, Mohammed, 2007, The End of Pseudo-Science: Essays Refuting False Scientific Theories Taught in Schools, Colleges, and Universities, iUniverse, 86 p.

Albrecht, Berhard, 2018, Sagnac interferometer: Vienna, Austria, Gruppe Arno Rauschenbeutel, ATI, TU Wien, Accessed 20181019 [https://go.glennborchardt.com/SagnacF32].

Alfvén, Hannes, 1977, Cosmology: Myth or science, *in* Yourgrau, Wolfgang, and Breck, A.D., eds., Cosmology, History, and Theology: New York, Plenum Press, p. 1-14. [http://go.glennborchardt.com/alfven77].

Almassi, Ben, 2009, Trust in expert testimony: Eddington's 1919 eclipse expedition and the British response to general relativity: Studies in History and Philosophy of Modern Physics, v. 40, p. 57-67. [http://go.glennborchardt.com/Almassi09].

Anon, 2005, Galaxy Clusters Formed Early, Accessed 20161027 [http://go.glennborchardt.com/younggalclus].

Anon, 2017, GW170817, Accessed 20171023 [http://go.glennborchardt.com/Wiki17LIGOGW170817].

Arp, Halton, 1998, Seeing red: Redshifts, cosmology and academic science: Montreal, Apeiron, 306 p.

Ashmore, Lyndon, 2006, Big Bang Blasted, Booksurge LLC, 306 p.

Bergamini, David, 1962, The universe: New York, Time, 192 p.

Bodanis, David, 2000, E=mc^2: A Biography of the World's Most Famous Equation: New York, Walker & Company, 337 p.

Bohm, David, 1957, Causality and Chance in Modern Physics: New York, Harper and Brothers, 170 p.

Bondi, Hermann, and Gold, Thomas, 1948, The steady-state theory of the expanding universe: Monthly Notices of the Royal Astronomical Society, v. 108, no. 3, p. 252-270.

Borchardt, Glenn, 1974, The SIMAN coefficient for similarity analysis: Classification Society Bulletin, v. 3, no. 2, p. 2-8 [http://go.glennborchardt.com/SIMAN].

Borchardt, Glenn, 1978, Catastrophe theory: Application to the Permian mass extinction: Comments and reply: Geology, v. 6, p. 453-454.

Borchardt, Glenn, 1984, The scientific worldview: Berkeley, California, Progressive Science Institute, 343 p. [http://doi.org/10.13140/RG.2.2.16123.52006].

Borchardt, Glenn, 2004, Ten assumptions of science and the demise of 'cosmogony', Proceedings of the Natural Philosophy Alliance, v. 1, no. 1, p. 3-6 [http://doi.org/10.13140/2.1.2638.5607].

Borchardt, Glenn, 2004, The ten assumptions of science: Toward a new scientific worldview: Lincoln, NE, iUniverse, 125 p. [Free download at http://doi.org/10.13140/RG.2.2.13320.21761].

Borchardt, Glenn, 2007, Infinite universe theory, Proceedings of the Natural Philosophy Alliance: Storrs, CN, v. 4, no. 1, p. 20-23 [http://doi.org/10.13140/RG.2.1.3515.0247].

Borchardt, Glenn, 2007, The Scientific Worldview: Beyond Newton and Einstein: Lincoln, NE, iUniverse, 411 p. [http://scientificphilosophy.com].

Borchardt, Glenn, 2008, Resolution of the SLT-order paradox, Proceedings of the Natural Philosophy Alliance: Albuquerque, NM, v. 5 [http://doi.org/10.13140/RG.2.1.1413.7768].

Borchardt, Glenn, 2009, The physical meaning of $E=mc^2$, Proceedings of the Natural Philosophy Alliance: Storrs, CN, v. 6, no. 1, p. 27-31 [http://doi.org/10.13140/RG.2.1.2387.4643].

Borchardt, Glenn, 2011, Einstein's most important philosophical error, in Volk, Greg, Proceedings of the Natural Philosophy Alliance, 18th Conference of the NPA, 6-9 July, 2011: College Park, MD, Natural Philosophy Alliance, Mt. Airy, MD, v. 8, p. 64-68 [http://doi.org/10.13140/RG.2.1.3436.0407].

Borchardt, Glenn, and Puetz, S.J., 2012, Neomechanical gravitation theory, in Volk, Greg, Proceedings of the Natural Philosophy Alliance, 19th Conference of the NPA, 25-28 July: Albuquerque, NM, Natural Philosophy Alliance, Mt. Airy, MD, v. 9, p. 53-58 [http://doi.org/10.13140/RG.2.1.3991.0483] (See also the oral presentation at http://go.glennborchardt.com/NGTvideo).

Brecher, Kenneth, 1977, Is the Speed of Light Independent of the Velocity of the Source?: Physical Review Letters, v. 39, no. 17, p. 1051-1054. [http://go.glennborchardt.com/Brecher77].

Bruno, Giordano, 1600, Giordano Bruno quotes. [http://go.glennborchardt.com/Brun].

Bryant, Steven, 2011, The twin paradox: Why it is required by relativity, *in* Volk, Greg, ed., in Proceedings of the Natural Philosophy Alliance, 18th Conference of the NPA, 6-9 July, 2011, College Park, MD, Natural Philosophy Alliance, Mt. Airy, MD [http://go.glennborchardt.com/Bryant11Twin].

Bryant, S.B., 2016, Disruptive: Rewriting the rules of physics: El Cerrito, CA, Infinite Circle Publishing, 312 p.

Bryant, Steven, and Borchardt, Glenn, 2011, Failure of the relativistic hypercone derivation, in Volk, Greg, Proceedings of the Natural Philosophy Alliance, 18th Conference of the NPA, 6-9 July: College Park, MD, Natural Philosophy Alliance, Mt. Airy, MD, v. 8, p. 99-101 [http://doi.org/10.13140/RG.2.1.1404.8406].

Calvin, John, 1536 [1945], The Institutes of the Christian Religion: Grand Rapids, MI, Christian Classics Ethereal Library, 1221 p. [http://go.glennborchardt.com/Cal].

Chappell, J.E., Jr., 1965, Georges Sagnac and the Discovery of the Ether: Archives Internat. d'Hist. des Sciences, v. 18, p. 177.

Chopra, Deepak, and Kafatos, M.C., 2017, You are the universe: Discovering Your Cosmic Self and Why It Matters, Harmony, 290 p.

Chou, C.W., and others, 2010, Optical clocks and relativity: Science, v. 329, no. 5999, p. 1630-1633. [http://doi.org/10.1126/science.1192720].

Collingwood, R.G., 1940, An Essay on Metaphysics: Oxford, Clarendon Press, 354 p.

Conselice, C.J., Wilkinson, Aaron, Duncan, Kenneth, and Mortlock, Alice, 2016, The Evolution of Galaxy Number Density at z > 8 and Its Implications: The Astrophysical Journal, v. 830, no. 2, p. 83. [http://doi.org/10.3847/0004-637X/830/2/83].

Davies, P.C.W., 1981, The edge of infinity: where the universe came from and how it will end: New York, Simon and Schuster, 194 p.

de Climont, Jean, 2012, The worldwide list of dissident scientists, 1700 p. [http://go.glennborchardt.com/deClimont12dissidentlist].

de Climont, Jean, 2016, The worldwide list of dissident scientists [http://go.glennborchardt.com/declimont16dissidentlist].

de Sitter, Willem, 1913, An Astronomical Proof for the Constancy of the Speed of Light (English translation): Physik. Zeitschr. 14, 429, (1913) v. 14, p. 429. [http://go.glennborchardt.com/desitter13light].

Descartes, Rene, 1644 [1991], Principles of philosophy: Boston, MA, Kluwer Academic, 324 p. [http://go.glennborchardt.com/Descartes1644].

Dinu, Ionel 2010, Rudiments of a theory of aether: General Science Journal, v. 2370, p. 1-47. [http://go.glennborchardt.com/Dinu10aether].

Disney, M.J., 2011, Doubts about Big Bang Cosmology, *in* Alfonso-Faus, Antonio, ed., Aspects of today's cosmology: Rijeka, Croatia, InTech, p. 1-11. [http://go.glennborchardt.com/Disney11doubtsBBT].

Dowdye, E.H., Jr., 2010, Findings convincingly show no direct interaction between gravitation and electromagnetism in empty vacuum space, *in* Volk, Greg, Proceedings of the Natural Philosophy Alliance, 17th Conference of the NPA, 23-26 June, 2010: Long Beach, CA, Natural Philosophy Alliance, Mt. Airy, MD, v. 7, p. 131-136 [http://go.glennborchardt.com/Dowdye10Findings].

Dowdye, E.H., Jr., 2012, Discourses & Mathematical Illustrations Pertaining to the Extinction Shift Principle Under the Electrodynamics of Galilean Transformations (3rd ed.), Dr. Edward Henry Dowdye, Jr., 114 p. [http://go.glennborchardt.com/Dowdye12Extinctionbook].

Dyson, F.W., Eddington, A.S., and Davidson, Charles, 1920, A Determination of the Deflection of Light by the Sun's Gravitational Field, from Observations Made at the Total Eclipse of May 29, 1919: Philosophical Transactions of the Royal Society A: Mathematical, Physical and Engineering Sciences, v. 220, no. 571-581, p. 291-333. [http://doi.org/10.1098/rsta.1920.0009].

Edwards, M.R., ed., 2002, Pushing gravity: New perspectives on Le Sage's theory of gravitation: Montreal, Apeiron, 316 p.

Einstein, Albert, 1905 [1923], On the electrodynamics of moving bodies, *in* Einstein, A., Lorentz, H.A., Weyl, H., and Minkowski, H., eds., The principle of relativity: New York, Dover, p. 37-65. [http://go.glennborchardt.com/AE05Ontheelectro].

Einstein, Albert, 1905 [1965], Concerning an Heuristic Point of View Toward the Emission and Transformation of Light: American Journal of Physics, v. 33, n. 5, May 1965, v. 33, no. 5, p. 1-16. [http://go.glennborchardt.com/Einstein05Concerning].

Einstein, Albert, 1913 [1993], Letter from Einstein to Freundlich, August 1913, *in* Klein, Martin J., Kox, A. J., and Schulman, Robert, eds., The Collected Papers of Albert Einstein, The Swiss Years: Correspondence, 1902-1914 Princeton, Princeton: Princeton University Press, p. 354 (Document 472). [http://go.glennborchardt.com/AElightnotclassic].

Einstein, Albert, 1916 [1923], The foundation of the general theory of relativity, *in* Einstein, A., Lorentz, H.A., Weyl, H., and Minkowski, H., eds., The principle of relativity: New York, Dover, p. 109-164. [http://go.glennborchardt.com/Einstein16theprinciple].

Einstein, Albert, 1916 [1923], Hamilton's principle and the general theory of relativity, *in* Einstein, A., Lorentz, H.A., Weyl, H., and Minkowski, H., eds., The principle of relativity: New York, Dover, p. 165-173. [http://go.glennborchardt.com/Einstein16theprinciple].

Einstein, Albert, 1917 [1923], Cosmological considerations of the general theory of relativity, in Einstein, A., Lorentz, H.A., Weyl, H., and Minkowski, H., eds., The principle of relativity: New York, Dover, p. 175-188. [http://go.glennborchardt.com/Einstein16theprinciple].

Einstein, Albert, 1920, Ether and the Theory of Relativity: An address delivered on May 5th, 1920, in the University of Leyden [http://go.glennborchardt.com/Einstein20recantation].

Engels, Frederick, 1883 [1925], Dialectics of nature: Moscow, Progress, 403 p. [http://go.glennborchardt.com/Engles83dialectics].

Falkenstein, Eric, 2010, Eddington's Experiment Was Bogus, Accessed 20130206 [http://go.glennborchardt.com/Falkenstein10eddingtonbogus].

Farmer, B.L., 1997, Universe alternatives: Emerging concepts of size, age, structure and behavior (2nd ed.): El Paso, TX, Billy L. Farmer, 129 p. [See my review at: http://go.glennborchardt.com/Farmer97universealternativesreview]

Feynman, Richard, Leighton, R.B., and Sands, Matthew, 1964, The Feynman lectures on physics, Addison Wesley, v. 1 [http://go.glennborchardt.com/Feynman64energy].

Fox, J. G., 1965, Evidence Against Emission Theories: American Journal of Physics, v. 33, no. 1, p. 1-17 [http://doi.org/10.1119/1.1971219].

Freeman, Morgan, 2011, Yves Couder Explains Wave/Particle Duality via Silicon Droplets: "Through the Wormhole", [http://go.glennborchardt.com/Freeman11YoutubeQM].

Friedmann, Alexander, 1922, Über die Krümmung des Raumes: Zeitschrift für Physik, v. 10, no. 1, p. 377–386. [http://doi.org/10.1007/BF01332580].

Frommert, Harmut, and Kronberg, Christine, 2005, The Milky Way Galaxy, Accessed 20171023 [http://go.glennborchardt.com/Frommert05MWstarnumber].

Galaev, Y.M., 2002, The measuring of ether-drift velocity and kinematic ether viscosity within optical waves band (English translation): Space-time & Substance, v. 3, no. 5, p. 207-224. [http://go.glennborchardt.com/Galaevaether].

Gamow, George, 1954, Modern cosmology: Scientific American, v. 190, no. 3, p. 55-63.

Gamow, George, 1961, The creation of the universe: New York, Viking, 147 p.

Gardi, Lori, 2017, A medium for the propagation of light revisited, in Restoring Critical Thinking in Science, University of British Columbia, Vancouver, Canada, John Chappell Natural Philosophy Society, Caldeon, MI, p. 74-78.

Gardner, Martin, 1962, Relativity for the Million: New York, Macmillan, p. 66.

Gary, Stuart, 2011, Astronomers find old heads in a young crowd, Accessed 20171022 [http://go.glennborchardt.com/Gary11elderlygalaxies].

Gott, J.R., III, Gunn, J.E., Schramm, D.N., and Tinsley, B.M., 1976 [1977], Will the universe expand forever?, *in* Gingerich, Owen, ed., Cosmology + 1: Readings from Scientific American: San Francisco, Freeman, p. 87-89.

Greiner, Walter, and Hamilton, Joseph, 1980, Is the vacuum really empty?: American Scientist, v. 68, no. 2, p. 154-164.

Hadhazy, Adam, 2010, Ancient City of Galaxies Looks Surprisingly Modern, Accessed 20171022 [http://go.glennborchardt.com/Hadhazy10elderlygalaxies].

Hafele, J.C., and Keating, R.E., 1972a, Around-the-World Atomic Clocks: Predicted Relativistic Time Gains: Science, v. 177, no. 4044, p. 166-168. [http://doi.org/10.1126/science.177.4044.166].

Hafele, J.C., and Keating, R.E., 1972b, Around-the-World Atomic Clocks: Observed Relativistic Time Gains: Science, v. 177, no. 4044, p. 168-170. [http://doi.org/10.1126/science.177.4044.168].

Harrington, J.D., and Clavin, Whitney, 2013, Planck Mission Brings Universe into Sharp Focus, Accessed 20171022 [http://go.glennborchardt.com/Harrington13focus].

Hawking, S.W., 1988, A brief history of time: From the big bang to black holes: New York, Bantam Books, 272 p.

Hawking, Stephen, and Mlodinow, Leonard, 2012, The grand design: New York, Bantam Books, 208 p.

Hawkins, G.S., 1962, Expansion of the universe: Nature, v. 194, p. 563-564.

Heilprin, John, and Borenstein, Seth, 2013, Scientists find possible hint of dark matter from cosmos, Taiwan News, Associated Press [http://go.glennborchardt.com/Heilprin13DM].

Holtzner, Steve, 2016, Using the Kinetic Energy Formula to Predict Air Molecule Speed, Physics I For Dummies: (2nd ed.), p. 408. [http://go.glennborchardt.com/Holtzer16airmoleculespeed].

Houlgate, Stephen, ed., 1998, The Hegel reader: Malden, MA, Blackwell Publishing, p. 270.

Hoyle, Fred, 1948, A new model for the expanding universe: Monthly Notices of the Royal Astronomical Society, v. 108, no. 3, p. 372-382.

Hubble, Edwin, 1929, A relation between distance and radial velocity among extra-galactic nebulae: Proceedings of the National Academy of Sciences, v. 15, no. 3, p. 168-173. [http://doi.org/10.1073/pnas.15.3.168]

Hubble, Edwin, 1936, Effects of Redshifts on the Distribution of Nebulae: Astrophysical Journal, v. 84, no. 12, p. 517-554.

Hubble, Edwin, 1947, The 200-inch telescope and some problems it may solve: Publications of the astronomical society of the Pacific, v. 59, no. 349, p. 153-167.

Hubble, Edwin, 1953, The Law of Red-shifts: George Darwin Lecture, delivered by Dr Edwin Hubble on 1953 May 8: Monthly Notices of the Royal Astronomical Society, v. 113, no. 6, p. 658-666 [http://doi.org/10.1093/mnras/113.6.658].

Huygens, Christiaan, 1690 [2005], Treatise on light: Chicago, IL, University of Chicago Press, 129 p. [http://go.glennborchardt.com/Huygens1690light].

Jastrow, Robert, 1978, God and the astronomers: New York, Norton, 136 p.

Jarrett, T.H., 2004, Large Scale Structure in the Local Universe: The 2MASS Galaxy Catalog: PASA, 21, 396, Accessed 20171022 [http://go.glennborchardt.com/Jarrett04universestructure].

Jourdain, P.E.B., 1913, The Principle of Least Action: Chicago, Open Court, 83 p.

Kant, Immanuel, 1755 [2009], Universal Natural History and Theory of the Heavens, *in* Ian Johnston, Translator, ed. Arlington, VA, Richer Resources Publications, p. 43. [http://go.glennborchardt.com/Kant1755worldsystem].

Kashlinsky, A., Atrio-Barandela, F., Ebeling, H., Edge, A., and Kocevski, D., 2010, A New Measurement of the Bulk Flow of X-Ray Luminous Clusters of Galaxies: The Astrophysical Journal Letters, v. 712, no. 1, p. L81-L85 [http://doi.org/10.1088/2041-8205/712/1/L81].

Kelly, A.G., 2000, Hafele & Keating Tests: Did They Prove Anything? [http://go.glennborchardt.com/KellyonHK].

Kemp, R.L., 2012, Super Principia Mathematica, Accessed 20171022 [http://go.glennborchardt.com/Kemp12aethersubspara6].

King, Bob, 2014, 9,096 Stars in the Sky — Is That All?, Sky and Telescope [http://go.glennborchardt.com/King14visiblestars].

Kluger, Jeffrey, 2012, The candidates: The Higgs Boson, Time, Inc., Accessed 20171022 [http://go.glennborchardt.com/Kluger12Higgs].

Kopeikin, S.M., and others, 2016, Chronometric measurement of orthometric height differences by means of atomic clocks: Gravitation and Cosmology, v. 22, no. 3, p. 234-244. [http://doi.org/10.1134/s0202289316030099].

Krauss, L.M., 2012, A universe from nothing: Why there is something rather than nothing: New York, Free Press, 224 p.

Kuhn, T.S., 1962, The Structure of Scientific Revolutions: Chicago, University of Chicago Press, 172 p.

Kuhn, T.S., 1996, The structure of scientific revolutions (3 ed.): Chicago, University of Chicago Press, 212 p.

Kuhn, T.S., 1962 [2012], The structure of scientific revolutions (With an Introductory Essay by Ian Hacking) (50th Anniversary ed.): Chicago; London, The University of Chicago Press, 264 p.

Laplace, P.S., 1816, Sur la vitesse du son dans l'air et dans l'eau: Ann. Chim. Phys., v. 3, p. 238-241.

Lemaître, Abbé Georges, 1950, The primeval atom: An essay on cosmogony: New York, D. Van Nostrand, 186 p.

Linde, Andrei, and Vanchurin, Vitaly, 2010, How many universes are in the multiverse?: Physical Review D, v. 81, no. 8, p. 083525. [http://doi.org/10.1103/PhysRevD.81.083525].

Lenin, V.I., 1908 [1927], Materialism and empirio-criticism: Critical comments on a reactionary philosophy: New York, International, 397 p. [http://go.glennborchardt.com/Lenin1908materialism].

Lerner, E.J., 1992, The Big Bang Never Happened: New York, Vintage Books, 496 p.

Lewis, Geraint, 2014, Grey is the new black hole: is Stephen Hawking right?: The Conversation APA citation:, Accessed 20171022 [http://go.glennborchardt.com/Lewis14BHaregrey].

Lewis, G.N., 1926, The conservation of photons: Nature, v. 118, no. 2981, p. 874-875. [http://go.glennborchardt.com/Lewis26photonorigin].

Lucretius, 60 BCE [1994], On the Nature of the Universe: New York, Penguin Classics, 336 p.

Magness, P.W., and Murphy, R.P., 2014, Challenging the Empirical Contribution of Thomas Piketty's Capital in the 21st Century: Journal of Private Enterprise, Spring 2015; GMU School of Public Policy Research Paper No. 15-2. [http://go.glennborchardt.com/Magness14antiPiketty].

Martin Jr., R.C., 1999, Astronomy on Trial: A Devastating and Complete Repudiation of the Big Bang Fiasco: Lanham, Maryland), UPA, 264 p.

Martínez, A.A., 2004, Ritz, Einstein, and the Emission Hypothesis: Physics in Perspective, v. 6, no. 1, p. 4-28 [http://doi.org/10.1007/s00016-003-0195-6].

Marmet, Paul, 1990, Big Bang Cosmology Meets an Astronomical Death: 21st Century, Science and Technology (P.O. Box, 17285, Washington, D.C. 20041), v. 3, no. 2, p. 52-59. [http://go.glennborchardt.com/Marmet90BBTdeath].

Maupertuis, P. L. M. de, 1751, Essai de cosmologie: Leiden, 238 p. [http://go.glennborchardt.com/Maupertuis1751Essai].

Michel, Jean-Baptiste, Shen, Yuan Kui, Aiden, Aviva Presser, Veres, Adrian, Gray, Matthew K., Pickett, Joseph P., Hoiberg, Dale, Clancy, Dan, Norvig, Peter, Orwant, Jon, Pinker, Steven, Nowak, Martin A., and Aiden, Erez Lieberman, 2011, Quantitative Analysis of Culture Using Millions of Digitized Books: Science, v. 331, no. 6014, p. 176-182. [10.1126/science.1199644].

McKenna, Josephine, 2014, Pope says evolution, Big Bang are real, USA Today [http://go.glennborchardt.com/McKenna14popesBBT].

McLeod, S.A., 2010, Sensorimotor Stage, Accessed 20171022 [http://go.glennborchardt.com/McLeod10piaget].

Michelson, A.A., and Morley, E.W., 1887, On the relative motion of the earth and the luminiferous ether: American Journal of Science, v. 39, p. 333-345. [http://go.glennborchardt.com/MMX1887].

Mitchel, W.C., 1994, The cult of the big bang: Was there a bang?: Carson City, NV, Common Sense Books, 240 p.

Mitchel, W.C., 2002, Bye bye big bang: Hello reality: Carson City, NV, Common Sense Books, 448 p.

Mitton, Simon, 2011, Fred Hoyle: A life in science: New York, NY, Cambridge University Press, 384 p.

Moody, Richard, Jr., 2009, The eclipse data from 1919: The greatest hoax in 20th century science, Accessed 20171029 [http://go.glennborchardt.com/Moody09eclipse].

Narliker, Jayant, 1981, Was there a big bang?: New Scientist, v. 91, p. 19-21.

NASA, 2014, Dark Energy, Dark Matter, Accessed 20171022 [http://go.glennborchardt.com/NASA14DM].

Newton, Isaac, 1687 [1846], Philosophiae Naturalis Principia Mathematica. Translated by Andrew Motte: New York, NY, Daniel Adee, 581 p. [http://go.glennborchardt.com/Newton1687Principia].

Newton, Isaac, 1688, Letter to Dr. Covel Feb. 21, (1688-9) Thirteen Letters from Sir Isaac Newton to J. Covel, D.D. (1848) [http://go.glennborchardt.com/Newton1688Covelletter].

Newton, Isaac, 1692 [1965], Fundamental principles of natural philosophy and four letters to Richard Bentley, *in* Munitz, M.K., ed., Theories of the universe: From Babylonian myth to modern science: New York, Free Press, p. 202-219.

Newton, Isaac, 1718, Opticks or, a treatise of the reflections, refractions, inflections and colours of light (Second ed.): London, Royal Society, 382 p. [http://go.glennborchardt.com/Newton1718Optics].

Newton, Isaac, 1726 [1999], Philosophiae Naturalis Principia Mathematica, General Scholium. Translated by I. Bernard Cohen and Anne Whitman (3rd ed.), University of California Press, 974 p.

Ost, Laura, 2010, NIST Clock Experiment Demonstrates That Your Head is Older Than Your Feet: NIST, (September 28, 2010 ed.), Accessed 20171114 [http://go.glennborchardt.com/Ostheadolderthanfeet].

Piketty, Thomas, 2014, Capital in the twenty-first century: Cambridge, MA, Harvard University Press, 685 p. [http://go.glennborchardt.com/Piketty13Capitalism].

Popper, K.R., 2002, The Logic of Scientific Discovery (15th ed.): New York, Routledge, 544 p.

Pound, R.V., and Rebka, G.A., 1960, Apparent Weight of Photons: Physical Review Letters, v. 4, no. 7, p. 337-341. [http://go.glennborchardt.com/Pound60weightofphotons].

Prokoph, Andreas, and Puetz, S.J., 2015, Period-Tripling and Fractal Features in Multi-Billion Year Geological Records: Mathematical Geosciences, p. 1-20. [http://doi.org/10.1007/s11004-015-9593-y].

Puetz, S.J., and Borchardt, Glenn, 2011, Universal Cycle Theory: Neomechanics of the Hierarchically Infinite Universe: Denver, Outskirts Press, 626 p. [*http://www.scientificphilosophy.com/*].

---, 2015, Quasi-periodic fractal patterns in geomagnetic reversals, geological activity, and astronomical events: Chaos, Solitons & Fractals, v. 81, no. Part A, p. 246–270. [http://doi.org/10.1016/j.chaos.2015.09.029].

Puetz, S.J., Prokoph, Andreas, and Borchardt, Glenn, 2016, Evaluating alternatives to the Milankovitch theory: Journal of Statistical Planning and Inference, v. 170, p. 158-165 [http://doi.org/10.1016/j.jspi.2015.10.006].

Puetz, S.J., Prokoph, Andreas, Borchardt, Glenn, and Mason, E.W., 2014, Evidence of synchronous, decadal to billion year cycles in geological, genetic, and astronomical events: Chaos, Solitons & Fractals, v. 62–63, no. 0, p. 55-75. [http://doi.org/10.1016/j.chaos.2014.04.001].

Rejkuba, M., Dubath, P., Minniti, D., and Meylan, G., 2008, Masses and M/L Ratios of Bright Globular Clusters in NGC 5128: Proceedings of the International Astronomical Union. IAU Symposium, v. 246, p. 418–422 [*http://doi.org/10.1017/S1743921308016074*].

Ricker, H.H., 2015, The origin of the equation $E=mc^2$, Accessed 20171022 [http://go.glennborchardt.com/Ricker15mc2origin].

Ritz, Walter, 1908, Recherches critiques sur l'Électrodynamique Générale (See also the English translation in the link): Annales de Chimie et de Physique, v. 13, p. 145-275. [http://go.glennborchardt.com/Ritz1908lightparticles].

Roach, John, 2007, "No Two Snowflakes the Same" Likely True, Research Reveals: National Geographic News. [http://go.glennborchardt.com/Roach07snowflakes].

Rodgers, Paul, 2014, Einstein Was Right: You Can Turn Energy Into Matter, Accessed 20171023 [http://go.glennborchardt.com/Rodgers14einsteinism].

Ross, Sydney, 1962, Scientist: The story of a word: Annals of Science, v. 18, no. 2, p. 65-85.

Rubin, Vera C., 2000, One Hundred Years of Rotating Galaxies: Publications of the Astronomical Society of the Pacific, v. 112, p. 747-750. [https://ned.ipac.caltech.edu/level5/Sept04/Rubin/paper.pdf].

Rubin, Vera C., and Ford, W.K., Jr., 1970, Rotation of the Andromeda Nebula from a Spectroscopic Survey of Emission Regions: Astrophysical Journal, v. 159, p. 379. [10.1086/150317].

Sagnac, Georges, 1913a, The Demonstration of the Luminiferous Aether by an Interferometer in Uniform Rotation: Comptes Rendus, v. 157, p. 708–710.

Sagnac, Georges, 1913b, On the Proof of the Reality of the Luminiferous Aether by the Experiment with a Rotating Interferometer: Comptes Rendus, v. 157, p. 1410–1413.

Sauvé, Vincent, 2016, Edwin Hubble... and the myth that he discovered an expanding universe, Accessed 20171022 [http://go.glennborchardt.com/Sauve07Hubble].

Schlafly, R.S., 2011, How Einstein ruined physics: Motion, symmetry, and revolution in science: Lexington, KY, CreateSpace Independent Publishing Platform, 350 p.

Schlegel, Richard, 1973, Phsyical Sciences: Flying Clocks and the Sagnac Effect: Nature, v. 242, p. 180. [http://doi.org/10.1038/242180a0].

Schroeder, Paul, 2006, The universe is otherwise, BookSurge Publishing, 198 p.

Shapiro, S.L., Stark, R.F., and Teukolsky, S.A., 1985, The Search for Gravitational Waves: American Scientist, v. 73, p. 248-257.

Shaw, Duncan W., 2012, The Cause of Gravity—a Concept: Physics Essays, v. 25, no. 1, p. 66-75 [*http://doi.org/10.4006/0836-1398-25.1.66*].

Shaw, Duncan W., 2014, Reconsidering Maxwell's Aether: Physics Essays, v. 27, no. 4, p. 601-607 [*http://doi.org/10.4006/0836-1398-27.4.601*].

Silk, Joseph, 1973, Cosmological theory: Science, v. 181, p. 1038-1039.

Silk, Joseph, 1980, The big bang: The creation and evolution of the universe: San Francisco, Freeman, 394 p.

Silk, Joseph, 2002, The big bang (3rd ed.): New York, Freeman, 480 p.

Sinclair, Upton, 1935 [1994], I, Candidate for Governor and How I Got Licked: Berkeley, University of California Press, 272 p.

Slipher, Vesto, 1913, The Radial Velocity of the Andromeda Nebula: Lowell Observatory Bulletin 58, v. 2, no. 8, p. 56-57.

Smith, K.T., 2016, How many galaxies are in the universe?: Science, v. 354, no. 6314, p. 844-844 [http://doi.org/10.1126/science.354.6314.844-a].

Smith-Strickland, Kiona, 2015, This is the Oldest Galaxy We've Found So Far, Accessed 20161022
[http://go.glennborchardt.com/Smith15oldestgalaxy].

Smolin, Lee, 2006, The Trouble with Physics: The Rise of String Theory, the Fall of a Science, and What Comes Next: New York, NY, Houghton Mifflin, 392 p.

Spencer, D.E., and Shama, Uma, 1996, A new interpretation of the Hafele-Keating experiment, Accessed 20171022
[http://go.glennborchardt.com/Spencer96HKfraud].

Sujak, Peter, 2017, Einstein's repudiation of his own theory of relativity, *in* De Hilster, David, De Hilster, Robert, and Percival, Nick, eds., in Restoring Critical Thinking in Science, University of British Columbia, Vancouver, Canada, John Chappell Natural Philosophy Society, Caldeon, MI, p. 186-190.

Tatsumoto, Mitsunobu, and Rosholt, J.N., 1970, Age of the Moon: An Isotopic Study of Uranium-Thorium-Lead Systematics of Lunar Samples: Science, v. 167, no. 3918, p. 461-463.
[http://doi.org/10.1126/science.167.3918.461]

Thacker, Jerrold, 2013, The Shapiro Effect: Why Light From Distant Galaxies Is Redshifted, Accessed 20171022
[http://go.glennborchardt.com/Thacker13Shapiro].

Thompson, Andrea, 2009, Most Distant Galaxy With Big Black Hole Discovered, Accessed 20171022
[http://go.glennborchardt.com/Thompson09elderlygalaxy].

Van Flandern, Tom, 1998, The speed of gravity - What the experiments say: Physics Letters A, v. 250, no. 1-3, p. 11.
[http://go.glennborchardt.com/VanFlandern98gravityspeed].

Vaudrevange, P.M, Starkman, G.D., Cornish, N.J., and Spergel, D. N., 2012, Constraints on the topology of the Universe: Extension to general geometries: Physical Review D, v. 86, no. 8, p. 083526.
[http://doi.org/10.1103/PhysRevD.86.083526].

Vergano, Dan, 2014, Hubble Reveals Universe's Oldest Galaxies, Accessed 20171022
[http://go.glennborchardt.com/Vergano14elderlygalaxies].

Wald, R.M., 1977, Space, time, and gravity: The theory of the big bang and black holes: Chicago, University of Chicago Press, 131 p.

Waller, John, 2002, Einstein's Luck: The Truth Behind Some of the Greatest Scientific Discoveries: Oxford; New York, Oxford University Press, 308 p.

Wanjek, Christopher, and Steigerwald, Bill, 2005, Most Distant Galaxy Cluster Shows Universe 'Grew Up Fast', NASA Goddard Space Flight Center, Accessed 20171022
[http://go.glennborchardt.com/Wanjek05elderlygalaxies].

Whittaker, Sir Edmund, 1910 [1951], A history of the theories of aether and electricity: The classical theories: New York, Harper Torchbooks, v. 1, 434 p. [http://go.glennborchardt.com/Whittaker1910aetherhist].

Whittaker, Sir Edmund, 1953, A history of the theories of aether and electricity: The modern theories, 1900-1926: New York, Harper and Brothers, v. 2, 319 p.
[http://go.glennborchardt.com/Whittaker53aethermodern].

Young, Thomas, 1804, I. The Bakerian Lecture. Experiments and calculations relative to physical optics: Philosophical Transactions of the Royal Society of London, v. 94, p. 1-16. [http://doi.org/10.1098/rstl.1804.0001].

Zipf, G.K., 1949, Human Behavior and the Principle of Least Effort: An Introduction to Human Ecology: Cambridge, MA, Addison-Wesley, 573 p.

Glossary

ABERRATION. The discrepancy between the detection of the location of an object and its actual location. For instance, light takes about 8 minutes to travel from the Sun to Earth. Because Earth is rotating, what you observe visually as the Sun's image is no longer at the location indicated by that image. It would be off by about (8 min/(24 hr×60 min/hr)) X 360° = 2°, which is about four times the width of the Moon on the sky. Gravitation has no aberration, so orbital calculations need to correct data from visual observations involving light transit times that vary with distance. For example, without this correction, it would be impossible to produce accurate predictions of the future positions of the planets.

AD HOC HYPOTHESIS. "In science and philosophy, ad hoc means the addition of extraneous hypotheses to a theory to save it from being falsified. Ad hoc hypotheses compensate for anomalies not anticipated by the theory in its unmodified form. Scientists are often skeptical of scientific theories that rely on frequent, unsupported adjustments to sustain them. Ad hoc hypotheses are often characteristic of pseudo-scientific subjects such as homeopathy."[276]

AETHER. The medium theoretically responsible for light transmission, gravitation, and the formation of baryonic matter. Its tiny particle size precludes direct observation, although its effects are well known. This spelling is from 14th century Latin. It is sometimes spelled in the 12th century Old French form as "ether." Unfortunately, I am guilty of using the "ether" spelling in other books and papers. I switched to "aether" for two reasons: 1) it avoids confusion with the organic compound and 2) its

---

[276] http://go.glennborchardt.com/adhoc

Latin form and its definition as the precursor to ordinary matter apparently has precedence with René Descartes (1596-1650) where it occurs on p. 140 of the Latin version of his 1644 "Principles of Philosophy." For more detail and speculation on this subject, see Chapter 16.2.

AETHER DECELERATION THEORY (ADT). Theory that gravitation is caused by collisions from aether particles, resulting in baryonic acceleration and aether deceleration per Newton's Second Law of Motion. Because of their deceased activity, decelerated aether particles tend to be entrained by baryonic matter. This produces a low-pressure "halo" that allows access by relatively undecelerated aether particles whose activity level (pressure) increases with distance from baryonic matter. See the details on ADT in this section.

ANTIREDUCTIONIST. One who rejects the simplification of complex subjects on philosophical grounds. With the universe being infinite, this particular affliction is always with us per uncertainty. Antireductionists seem to feel that any particular statement about anything leaves out important details. They are known for a disease I call "yesbutitis." Probably in the interest of dialog, no matter what statement you make, they will have another, which says: "yes but, what about this or that other factor?" Of course, many complaints against classical mechanics and the philosophy of mechanism were valid. There never are a finite number of causes for a particular effect, so the mechanists can never list them all and the antireductionist will wait in vain for them to do so. Besides, antireductionists of the indeterministic type have another goal: the reduction of all things, not to matter, but to spirit.

BARYONIC MATTER. Ordinary matter, which includes electrons, positrons, atoms, and molecules that comprise the things of everyday existence. In general, baryonic matter can be sensed by the five senses directly or indirectly through instruments, as opposed to aether, which is matter too, but is much smaller and not so easily sensed.

BLUESHIFT. The process by which wavelength becomes shorter and/or frequency increases. The "blue" in the term implies a decrease in wavelength derived from the fact that high-frequency blue light has a shorter wavelength than low-frequency red light. It is a bit of a misnomer because color is determined by frequency, not wavelength. When wave velocity is constant, the relationship: $v = \lambda f$ remains constant, where v = velocity, m/s, λ = wavelength, m, and f = frequency, cycles/s.

CARRYING CAPACITY. In biology, carrying capacity generally occurs when a species accommodates to available resources. As carrying capacity is reached, it becomes increasingly difficult to obtain the necessities for life. For instance, if your life depended only on your obtaining gold from now-deficient streams, you probably would not last long. Homo sapiens has not yet reached carrying capacity, although there are many isolated instances where tribes and nations have reached population levels that were difficult to sustain without war and migration.

CONSUPPONIBLE. A description of multiple suppositions that are assumed to be true and without significant contradiction. The word appears to have been invented by R.G. Collingwood, a historian, who devised a method to discover fundamental assumptions.[277] The concept of consupponibility seems to give folks trouble. Maybe it is because they are accustomed to scores of contradictory ideas, such as walking on water, virgin birth, and living after dying. Modern physicists, in particular, are accustomed to wave-particle duality, the twin paradox, massless particles, and the explosion of the universe out of nothing. It is a

---

[277] Collingwood, 1940, An essay on metaphysics.

mess. These serve as justification for indeterministic claims that "the world is too complicated and too contradictory for us ordinary mortals to understand." An example of consupponibility would be these two (non-fundamental) assumptions: 1) that you have height and 2) that you have weight. These statements do not contradict one another. They are consupponible.

COSMOGONISTS. Those who assume that the universe had an origin. See the details below:

COSMOGONY. The study of the origin of the universe. Obviously, it assumes that the universe is finite and therefore actually had an origin. Creation myths, like the one in Genesis and in Big Bang Theory, have billions of supporters who remain steady in their beliefs. According to Google Books, the term became increasingly popular after 1710, with its popularity reaching a maximum in the 1880's and declining ever since, with only a few peaks in 1928 and thereafter (Figure 56). Today, the term is almost never used in cosmology, probably because it is an overt admission of assumptive bias, although that was not a problem for the priest who invented the Big Bang Theory.[278] Note that the alternative spelling, cosmogeny, is seldom used.

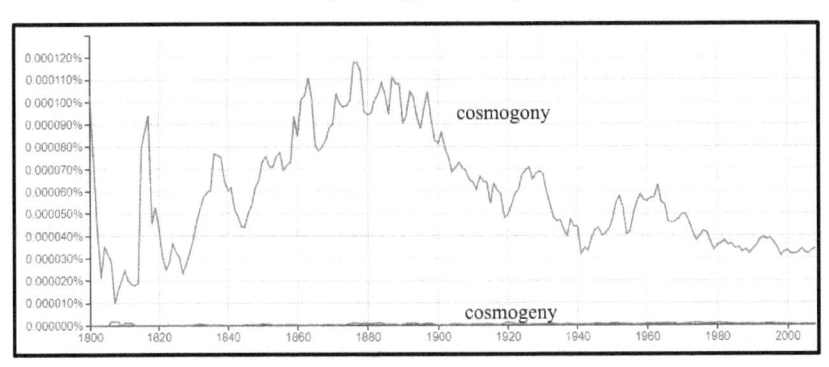

Figure 57 Ngram for the words "cosmogony" and "cosmogeny".[279]

---

[278] Lemaître, 1950, The primeval atom: An essay on cosmogony.
[279] Michel and others, ibid (accessed on 20181018).

The Big Bang Theory got popular for the same reason that relativity did: appeal to indeterminists. In the battle with science, the Big Bang Theory represents the last compromise, a final stand for cosmogony. The timing is a bit off for the extremely conservative, but even the Pope now accepts it.[280] The mostly religious representatives in congress would rather designate dollars for studies close to traditional cosmogony than to those opposed. As with relativity and most any theory or political action, understanding improves by using the old cliché: Follow the money.

COSMOLOGY. The study of the universe.

DARK ENERGY. A bogus calculation used to explain the motion required for the universe to expand per the Big Bang Theory. As of 20171123, Wikipedia put it this way "...**dark energy** is an unknown form of energy which is hypothesized to permeate all of space, tending to accelerate the expansion of the universe." As in the rest of regressive physics, this form of energy has been objectified in an extraordinarily clear way: "...dark energy contributes 68.3% of the total energy in the present-day observable universe. The mass–energy of dark matter and ordinary matter (baryonic) matter contribute 26.8% and 4.9%, respectively..."[281] In other words, dark energy is hereby considered a "constituent" of the universe, which it is not.

DARK MATTER. Decelerated aether surrounding all baryonic matter. Dark matter forms when high-velocity aether particles collide with baryonic matter during gravitation. Like the atmosphere, dark matter becomes entrained around cosmic bodies, forming an "aetherosphere" containing high-density aether having reduced aethereal pressure. Because aether

---

[280] McKenna, 2014, Pope says evolution, Big Bang are real.
[281] http://go.glennborchardt.com/darkenergy

particles have mass, cosmic bodies have three to ten times more mass than their illuminated constituents would indicate.

DETERMINISM. The assumption that there are material causes for all effects. The philosophical foundation for the legal system and most scientific disciplines other than physics and cosmogony. It is opposed to the idea of free will, which assumes that some effects are uncaused.

DETERMINISM-INDETERMINISM PHILOSOPHICAL STRUGGLE. The interminable conflict between the two primary ways of viewing the universe. The conflict begins at birth, getting in full swing as infants begin to learn that the universe consists of matter in motion. Eventually, they learn that most of those things and their motions occur independently of them or their thoughts about them. To survive, we need to learn about how the universe works. We must manipulate our environment to obtain food, shelter, and clothing. We rearrange the material things around us, discovering that there are material causes for effects—we tend to become "determinists." This learning process continues throughout life with varying degrees of success. Often, we cannot determine the material cause for a particular effect—who can see air or aether? Without an abiding faith that there always must be a material cause nevertheless, we might have a tendency to become "indeterminists." Even so, those who gain wide experience in successfully manipulating their environments, such as scientists, engineers, farmers, construction workers, and those in other hands-on occupations may have a tendency to generalize what they have learned. They might optimistically conclude that "there are material causes for all effects," becoming radical determinists. Others might pessimistically conclude just the opposite; becoming radical indeterminists in spite of whatever interactions they have had with the real world.

Science tends to favor determinism and religion tends to favor indeterminism. One part of the struggle involves denial that there is even a struggle. A complete, final proof of either position is impossible even though there is an enormous amount of data in support of determinism and none in support of indeterminism. That is because there is no way that the causes for all effects could ever be determined. Failure to find a cause for a particular effect allows indeterminists to assume that there isn't one. Of course, scientists must assume determinism, at least for the specialty they are working in. Religious folks must assume indeterminism as support for their belief in free will and the possibility of surviving as a non-physical entity after dying. Both opposing philosophical positions are highly ingrained, and would remain so except for one thing: Humans, being highly curious and capable, continue to broaden their experience with the external world. Again, determinism allows us to manipulate our surroundings, increasing life spans and the enjoyment thereof. Indeterminism, being a failed assumption with no data in support, has added nothing to our understanding and enjoyment of the world around us.

DIALECTICAL. Concerned with or acting through opposing forces.

DOPPLER EFFECT. The shortening or lengthening of wave motion due to a change in the distance between the source and the observer. The Doppler Effect is observed whenever a train passes by as it whistles. When it is coming toward us, the pitch is high (short wavelengths); when it is going away from us, the pitch is low (long wavelengths).

EINSTEINISM. "A statement or prediction that is true, but for the wrong reason."[282] Other, less preferred definitions are: 1) "a

---

[282] http://go.glennborchardt.com/Borchardt14massenergy

joke that becomes much less funny if it requires an explanation."[283] 2) "the perturbation of language or perception in order to put a positive spin on some aspect of Einstein's life. It may include distortion, omission, falsification, or corruption of the historic record in order to promote Einstein."[284] The most disparaging view was presented by Schlafly: "It is all a myth. Einstein did not invent relativity or most of the other things for which he is credited. He is mainly famous for popularizing the discoveries of others. We have all been duped."[285]

EINSTEIN'S RECANTATION. In an address delivered on May 5th, 1920, in the University of Leyden, Einstein said this:

Careful reflection teaches us that special relativity does not compel us to deny ether. We may assume its existence but not ascribe a definite state of motion to it ... There is a weighty reason in favour of ether. To deny ether is to ultimately assume that empty space has no physical qualities whatever.[286]

Do not get too excited about this. Here is the last sentence of his speech:

But this ether may not be thought of as endowed with the quality characteristic of ponderable media, as consisting of parts which may be tracked through time. The idea of motion may not be applied to it.

Sujak claims Einstein had over five papers from 1920 to 1934 in which he considered aether so obvious that he "openly repudiated his Special and General theories of relativity."[287]

---

[283] Attributed to Ian James Hay at http://go.glennborchardt.com/einsteinism
[284] Moody, 2009, The eclipse data from 1919: The greatest hoax in 20th century science.
[285] Schlafly, 2011, How Einstein ruined physics.
[286] Einstein, 1920, Ether and the Theory of Relativity.
[287] Sujak, 2017, Einstein's repudiation of his own theory of relativity.

ENERGY. A calculation used to describe the motion of matter or the potential for matter to move. Kinetic energy describes the unidirectional effect produced by a moving microcosm as it comes to a complete stop during a collision with an unyielding portion of the macrocosm ($E=1/2\ mv^2$). Potential energy describes the unidirectional effect produced by a microcosm after an unyielding portion of the macrocosm is removed (e.g., for gravitation, the equation would be $E=mgh$). Kinetic energy released during an explosion is bidirectional, producing motion in the forward and backward directions, so the calculation yields twice the unidirectional value ($E=mv^2$). If the motion is transferred bidirectionally to or from the aether, the equation is $E=mc^2$, which was first derived by Maxwell in 1862:

The derivation of $E=mc^2$ originates from Maxwell's formula [$f = \delta E/c\delta t$] which equates the force exerted on an absorbing body at the rate energy is received by the body. Since force is also the rate of the change of momentum of the body, which, by the conservation of momentum, is also the rate of change in the momentum of the radiation, the momentum lost by the radiation is equal to $1/c$ times the energy delivered to the body, or $M = E/c$. If the momentum of the radiation of a mass is M times the velocity c of the radiation, the equation $m = E/c^2$ is derived.[288]

Actually, this general relationship was supported by many earlier investigators, starting with Newton (see the Newton-Laplace Equation).

ENTROPY. The degree of divergence that exists among the submicrocosms within a microcosm. The opposite is negentropy, which is the degree of convergence that exists among the submicrocosms within a microcosm. Entropy increases when

---

[288] Ricker, 2015, The Origin of the Equation $E=mc^2$ [The true author of this quote is unclear. It was not Ricker. More info at: http://go.glennborchardt.com/emc2origin].

things go out of existence and decreases when things come into existence. This is discussed in more detail in the section on the Second Law of Thermodynamics and complementarity.

ETHER. I now use this spelling when referring to the imagined fixed medium once thought by indeterminists to be responsible for light transmission.[289] Its existence was disproven by the Michelson-Morley Experiment.[290] Earth travels around the Sun at a velocity of 30 km/s. If the ether was fixed, as assumed, then Earth's motion relative to that fixed medium would have been 30 km/s. It was much less than that, proving that the fixed ether did not exist. In other words, the fixed "ether" was falsified, but the "aether" consisting of particles in motion was not.

EXISTENCE. The xyz portion of the universe occupied by a microcosm after its formation via submicrocosmic convergence and before its destruction via submicrocosmic divergence.

FINITE PARTICLE THEORY (FPT). Indeterministic theory based on the assumption of microcosmic finity. Proponents tend to be absolutists who believe that the ideal end members of the solid matter-empty space continuum actually exist. The earliest version was proposed by the Greek atomists, who thought that matter consisted of perfectly identical little balls filled with "solid matter," in opposition to Aristotle's view that matter was infinitely subdividable. Accelerator experiments have supported Aristotle, but there still are believers in FPT who assume that a solid subatomic particle eventually will be found.[291] In this light, the Higgs boson has been anointed by the mass media as the

---

[289] As defined by Farmer, 1997, Universe alternatives. The word I now use for the medium for light transmission, gravitation, and the formation of baryonic matter is aether, which is not fixed.
[290] Michelson, A.A., and Morley, E.W., 1887, On the relative motion of the earth and the luminiferous ether. [Often referred to as "MMX."]
[291] Abu-Bakr, 2007, The End of Pseudo-Science.

"god particle." It supposedly is responsible for mass, although, ironically, it exists outside, not inside ordinary matter.

FORCE. A calculation used to describe the results of a collision of one microcosm with another per Newton's Second Law of Motion (F=ma, where m=mass and a=acceleration).

GENERAL RELATIVITY THEORY (GRT). The 1916 follow-up to Einstein's Special Relativity Theory[292] that continued his 1905 objectification of time, combining it with matter to propose space-time as a 4-dimensional object. GRT is the essential foundation of the Big Bang Theory. Like Special Relativity Theory, it includes numerous ad hoc assumptions that violate earlier laws of physics that form essential parts of classical mechanics.

GRAVITATION. The tendency for microcosms to be pushed toward less active, more massive microcosms.

IMMATERIALISM. The solipsistic belief that the universe does not consist of matter in motion. In opposition to materialism, immaterialism assumes that the universe is an illusion or internal perception that would not exist without a perceiving being.

IMPERIMENT. A thought "experiment." I invented this as a proper replacement for what was formerly considered a "thought experiment" by immaterialists such as Einstein. Strictly speaking, an experiment only can occur outside the mind per the prefix "ex." Science discovers truth through observation and experiment. Imperiments may be useful for predicting experimental results, but they have no credence among

---

[292] Einstein, 1916, The foundation of the general theory of relativity; Einstein, 1916, Hamilton's principle and the general theory of relativity; Einstein, 1917, Cosmological considerations of the general theory of relativity.

materialists (scientists) until those experiments actually are performed.

INDETERMINISM. The assumption that some effects may not have material causes (e.g., ESP, ghosts, gods, creation, free will). Specifically, indeterminism is founded on fundamental assumptions that lead to the false conclusion that free will is possible. Attempts to provide scientific support for indeterminism are inherently in contradiction. Science only deals with matter in motion; imagined "things" that do not exist cannot be tested because they do not exist.

INFINITE UNIVERSE THEORY (IUT). Theory proposed in this book that the universe is infinite both microcosmically and macrocosmically. Founded on "The Ten Assumptions of Science," IUT assumes that nonexistence is impossible and that the universe had no beginning and will have no end. Among other numerous details, IUT requires a rejuvenation of aether as the medium for light and gravitation. Although I have been working on it since 1978, my version of IUT was formally proposed in a short paper in 2007.[293]

INTEGRABLE. The ability of microcosms to combine with other microcosms to form still larger microcosms. This is obvious for atoms, solar systems, galaxies, galaxy clusters and superclusters, which illustrate the hierarchical nature of the universe.[294] In math, it refers to a function or differential equation capable of being integrated, that is, brought together.[295]

LIGHT. Light is a wave in the aether. The universe displays only two phenomena: matter and the motion of matter. The idea that light could be matter and motion at the same time is one of the great paradoxes of regressive physics. Although it makes no

---

[293] Borchardt, 2007, Infinite universe theory.
[294] Puetz and Borchardt, 2011, Universal cycle theory.
[295] http://go.glennborchardt.com/integrate

sense, indeterminists sell it as if it were some special concoction that we lesser mortals would never understand. Certainly, we could never be as "smart as Einstein," who was forced to invent the indeterministic wave-particle concept after denying the deterministic aether concept. Waves occur as group phenomena in a sea of particles. In the case of light motion, when one of those particles interacts with matter, it produces a collision known as the photoelectric effect,[296] a phenomenon described by Einstein, for which he received his only Nobel Prize. This was proof that particles are involved in light transmission in the same way that nitrogen molecules are involved in sound transmission.

Unfortunately, Occam's razor failed to work in this case. Instead of considering this as a simple proof of the particulate nature of aether, Einstein took another route. Like so many others, he misinterpreted the Michelson and Morley[297] experiment as evidence proving the absence of aether as the medium for light transmission. This presented a slew of problems, because light obviously was transmitted as wave motion, which always requires a medium. He ended up combining the particulate nature and the wave nature in a single microcosm, a "wave packet" later called the "photon." It may seem silly to the uninitiated, but to this day, light supposedly travels as a particle through empty space, bringing its waves along with it. The image gets particularly ridiculous when one realizes that the wavelengths of certain types of electromagnetic radiation waves are more than a kilometer long.

---

[296] Einstein, 1905, Concerning an Heuristic Point.

[297] Michelson and Morley, 1887, On the relative motion of the earth and the luminiferous ether. [Their experiments actually never gave a null result. They expected to obtain 30 km/s, which was the known velocity of Earth as it travels around the Sun. Instead, according to Bryant, 2016, Disruptive (p. 233), their calculations yielded 8.06+0.66 km/s, with a less than 0.1% chance of being zero.]

Early on, Einstein was an "aether denier"—that is what made Special Relativity Theory famous. Nonetheless, the wiser, older Einstein recanted in 1920, saying: "Careful reflection teaches us that special relativity does not compel us to deny ether."

That also should have put the kibosh on the corpuscular theory of light. It did not.

LORENTZ CORRECTION FACTOR, γ. An equation derived by physicist Hendrik Lorentz, which is used throughout the mathematics of relativity:

$$\gamma = \frac{1}{\sqrt{1-v^2/c^2}}$$

Unlike others, I call it a "correction" factor, because that is all it is. Wave motion through a medium takes time. For instance, when source and detector are moving apart, the path of that motion is stretched out: it takes longer to detect that motion. Wikipedia plainly shows it to be nothing but an uncomplicated *measurement* problem:[298]

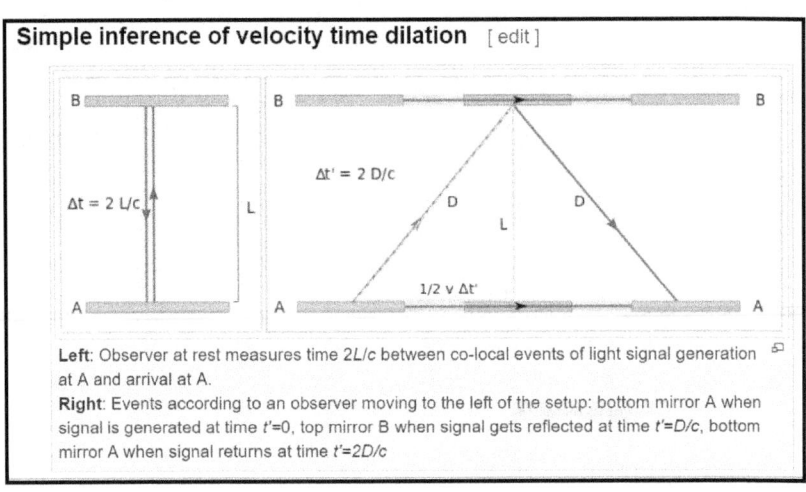

Simple inference of velocity time dilation [edit]
(diagram: Left shows vertical light path between mirrors A and B with $\Delta t = 2L/c$. Right shows triangular light path with $\Delta t' = 2D/c$ and horizontal displacement $1/2\, v\, \Delta t'$.)
Left: Observer at rest measures time 2L/c between co-local events of light signal generation at A and arrival at A.
Right: Events according to an observer moving to the left of the setup: bottom mirror A when signal is generated at time $t'=0$, top mirror B when signal gets reflected at time $t'=D/c$, bottom mirror A when signal returns at time $t'=2D/c$

---

[298] http://go.glennborchardt.com/TimeDilation

The total time for the light pulse to trace its path is given by

$$\Delta t' = \frac{2D}{c}.$$

The length of the half path can be calculated as a function of known quantities as

$$D = \sqrt{\left(\frac{1}{2}v\Delta t'\right)^2 + L^2}.$$

Elimination of the variables D and L from these three equations results in

$$\Delta t' = \frac{\Delta t}{\sqrt{1 - \frac{v^2}{c^2}}},$$

which expresses the fact that the moving observer's period of the clock $\Delta t'$ is longer than the period Δt in the frame of the clock itself.

In addition to the velocity of light, the "c" in the equation can be replaced by the velocity inherent to wave motion in any particular medium. For instance, sound in the atmosphere at sea level usually travels at 343 m/s at 20°C. In all media, changes in the travel path between source and detector produce the Doppler Effect. The phenomena that regressives call "time dilation" occurs when source or detector move away from each other and I suppose they might call it "time contraction" when they move toward each other. In the first instance, one must multiply by γ to calculate the increased length of the wave emitted from the source. In the second, one must divide by γ to calculate the decreased length of the wave emitted from the source. Note also that if c were infinite, γ would be 1 and there would be no effect.

As seen from the above, the Lorentz correction factor, γ, only concerns *measurement*. Strictly speaking, it has nothing to do with reality and everything to do with how we measure that reality. Time does not really dilate; mass does not really increase; length does not really contract as a result of motion through perfectly empty space. Claims to the contrary, such as those engendered from indeterministic interpretations of Special Relativity Theory, have no merit aside from unrecognized univironmental interactions that may occur in a special milieu that is by no means perfectly empty.

MACROCOSM. The environment of a microcosm. Strictly speaking, the macrocosm contains the rest of the infinite universe. Practically speaking, only the nearby portions of the universe generally have much influence on a particular microcosm.

MASS. Resistance to acceleration. We determine the mass of a microcosm by accelerating it with another microcosm of known mass and velocity. Mass and velocity are relative, so we have established standards, which on Earth, are relative to the acceleration of gravity (9.81 m/s^2). Because length and time also are relative, we have established conventions for those too. Realize, however, that these conventions are not absolute. Because the universe is infinite, measurements for each of them have a plus or minus and each tends to change over time. That is why we occasionally add a leap second to the length of the day as Earth's rotation rate slows. Although related, mass and matter are not identical. Mass is dependent on the internal motion of submicrocosms, which increases with the absorption of motion and decreases with the emission of motion per the $E=mc^2$ equation.[299]

MATTER. An abstraction for all things in existence. Above all, matter always contains other things within and without, *ad infinitum*. There are two basic types of matter: baryonic and aether. Although baryonic matter is what we ordinarily observe, aether is tiny and normally not directly detectable. Both have mass produced by constituents subject to interactions demonstrated by the $E=mc^2$ equation. The "solid matter" of the idealist does not exist.

---

[299] Borchardt, 2009, The physical meaning of $E=mc^2$.

MATERIALISM. The assumption that the external world exists after the observer does not, and that the universe consists of matter. It is the First Assumption of Science.

MATTER-MOTION TERMS. Valuable calculations in physics that generally multiply terms for matter and terms for motion. These describe neither matter nor motion, although they commonly have been objectified in regressive physics. For more details, see the chapter on progressive physics.

MATTER-SPACE CONTINUUM. A range or series of microcosms that are slightly different from each other and that exist between what we imagine to be perfectly solid matter and perfectly empty space.[300] Like all idealizations, solid matter and empty space do not and cannot exist.

The matter end member:

As mentioned, matter is an abstraction; there is no such thing as matter per se—there are only individual, unique examples of matter. The idea that solid matter must exist deep down at some level is still just that, an idea, ideal, or idealization that never occurs in nature. The Greek atomists imagined that atoms were true elementary particles filled with solid matter. The things we now call atoms appear to contain mostly empty space. Even so, some absolutists assume that we just have not gone far enough and that the nirvana of perfect solidity is theoretically possible.[301] At one time, the space between you and I may have been considered empty. Now we know that is not the case, for space is just the stuff that yields to the motion of other stuff. These ideals exist only in our brains—they help us understand the properties of various kinds of matter, but they can have no real existence. We use them to understand the intervening reality. It is good

---

[300] Modified from Merriam-Webster: http://go.glennborchardt.com/continuum
[301] Abu-Bakr, 2007, The End of Pseudo-Science.

enough for finding a doorway instead of a wall, even though the doorway contains matter in the form of air and the wall contains space. In Infinite Universe Theory, what we consider solid matter is simply a portion of the universe that offers more resistance to acceleration than other portions we consider empty space.

The space end member:

The absolutist's belief in the ideals of perfectly empty space, nothing, and nonexistence comes right out of the cosmogonical handbook whose precursors are the sacred texts of traditional religion. To insist, like the young Einstein and his positivist friends, that space is perfectly empty or immaterial makes one a rank idealist. To insist, as indeterminists are wont to do, that idealities could be or must be realities merely provides another roadblock to the ultimate acceptance of Infinite Universe Theory.

MECHANICS. The study of the universe in terms of matter and the motion of matter.

MECHANISM. The philosophy based on mechanics, which asserts that the universe has only two phenomena: matter and the motion of matter. The term also is used more widely to describe the material interactions that produce a particular effect.

MECHANIST. One who assumes that the universe consists only of matter in motion.

MICROCOSM. An xyz portion of the universe surrounded by an equally important environment called a macrocosm. Note that in conventional science microcosms are referred to as systems, which generally are considered more important than the environments in which they exist. In Infinite Universe Theory, microcosms cannot exist without their equally important macrocosms. Regardless of the immensity of a microcosm, in an Infinite Universe an infinitely large macrocosm still surrounds it.

The boundaries of a system sometimes are obvious: An apple, for instance, has a skin that roughly distinguishes it from its surroundings. At other times, the boundaries are not so obvious: A bee colony, for instance, has rather obscure boundaries when many of its members are far afield gathering nectar. Boundary selection is often difficult, always important, and frequently arbitrary. As scientists, we try to reduce arbitrariness by recording the location of boundaries with as much accuracy as possible. Our designation of a particular xyz portion of the universe as a microcosm faces the same problems, although in that instance, we treat its environment (the associated macrocosm) as equally important. Also, by attempting to treat the microcosm and the macrocosm equally, we are not as likely to miss important factors, as we would if we were biased toward one or the other.

MOMENTUM. A calculation describing the motion of matter. The equation is $P=mv$, where m=mass and v=velocity. As a matter-motion term, momentum neither exists nor occurs. What *does* exist is the matter described by its mass and what *does* occur is the motion described by the velocity of that matter relative to some other thing.

MOTION. An abstraction for all occurrences.

MULTIVERSE. A reformist hypothesis that there are other universes in addition to the one we can partially observe. Of course, the word itself is oxymoronic—there can be only one "uni" verse. The "universes" within the multiverse or "meta-universe" are called "parallel universes," "alternative universes," or "other universes." Like many reform attempts, this one has some hopeful aspects. It shows that cosmogonists are taking some of the first small steps out of the Big Bang Theory box.

According to Wikipedia[302] on 20170114, proponents of one of the multiverse hypotheses include Stephen Hawking,[16] Brian Greene,[17][18] Max Tegmark,[19] Alan Guth,[20] Andrei Linde,[21] Michio Kaku,[22] David Deutsch,[23] Leonard Susskind,[24] Alexander Vilenkin,[25] Yasunori Nomura,[26] Raj Pathria,[27] Laura Mersini-Houghton,[28][29] Neil deGrasse Tyson,[30] and Sean Carroll.[31] About the most ridiculous paper on the subject involved a calculation that there could be at least $10^{10^{10^7}}$ universes within the multiverse.[303]

NEOMECHANICAL GRAVITATION THEORY (NGT). Our proposal that gravitation was produced by local pressure differences in aether displaced by baryonic matter through vortex formation.[304] It has been modified considerably in the present book in the form of Aether Deceleration Theory.

NEOMECHANICS. Classical mechanics with the addition of infinity and its consupponible assumptions.

NEWTON-LAPLACE EQUATION. Early version of the $E=mc^2$ equation developed by Newton in 1687[305] and modified for media by Laplace.[306] Newton appears to be the first to recognize the relationship between mass and what eventually became known as "energy." Here is a summary of his ideas:

> He regarded light as, consisting of small material particles emitted from shining substances. He thought that these particles could be re-combined into solid matter, so that "gross bodies and light were convertible into one another;" that the particles of light and the particles of solid bodies acted mutually upon each other; those of light agitating and

---

[302] http://go.glennborchardt.com/multiverse [References in the Wikipedia article.]
[303] Linde and Vanchurin, 2010, How many universes are in the multiverse?
[304] Borchardt and Puetz, 2012, Neomechanical gravitation theory.
[305] Newton, 1687, Principia.
[306] Laplace, 1816, Sur la vitesse.

heating those of solid bodies, and the latter attracting and repelling the former.[307]

Note the similarity to the E=mc² equation in that Newton was implying that bodies subject to the Sun's rays would become internally agitated and heated. Their mass would increase in the same way it would if light were a particle that penetrated and resided in the object. Laplace later put that idea in an equation:

$$c = \sqrt{\frac{K}{\rho}}$$

Where:

c = velocity of light, m/s

K = bulk modulus, N/m² [Stiffness coefficient, a measure of energy.]

ρ = density, g/cm³ [A measure of mass.]

Probably because light is not supposed to require a medium for its transmission, this alternative equation for the relationship between energy and mass normally is missing in discussions of relativity. Fortunately, it is becoming more popular along with the increasing acceptance of aether.[308] It has broad utility for understanding wave motion through all media. When applied to the aether medium, it implies that increases in aether density result in decreases in light speed in tune with Chapter 16.3.[309]

---

[307] Ibid. From the foreword "Life of Sir Isaac Newton" written by N.W. Chittenden, p. 25.
[308] Gardi, Lori, 2017, A medium for the propagation of light revisited. [Lori gives an excellent description of the Newton-Laplace equation with respect to sound.]
[309] Thanks to Jesse Witwer for bringing the importance of the Newton-Laplace equation to my attention.

NGRAM. A word invented by Google that describes a plot showing the relative popularity of a particular word in Google books over time.[310]

OBJECTIFICATION. The consideration of motion as an object. Humans have objectified motion throughout history. Today, objectification is one of the primary roadblocks to understanding Infinite Universe Theory. Of the two fundamental phenomena, matter often comes to mind before motion. You can see matter, but you cannot see motion as a thing apart from matter. As we have seen, Einstein's relativity reeks with this old-fashioned tendency to objectify motion.[311] His "corpuscular" theory of light has a historic parallel with the "caloric fluid" theory of heat. Antoine Lavoisier (1743-1794),[312] the father of chemistry, imagined that "caloric fluid" traveled from object-to-object as a material entity. When you touched a burning ember, the caloric fluid supposedly flowed into your finger. Of course, what was once considered a "thing" is now correctly considered "motion," the vibration of things. In other words, heat is motion, not matter. If you wait too long to remove your finger the vibrations in the ember stimulate vibration in your finger, causing your discomfort and serious chemical transformations in your skin. Strictly speaking, heat does not exist; it occurs. In this case, the only thing that exists is the vibrating molecule. The molecule takes up three dimensions, while its motion does not.

Physicists have long abandoned the notion that heat is matter, although they have not done so for light. Today, few would ask whether heat had mass, since even today's most regressive physicists must regard heat as motion. Likewise, waves do not have mass, although the medium through which they travel does.

---

[310] http://go.glennborchardt.com/ngram; Michel and others, 2010, Quantitative Analysis of Culture Using Millions of Digitized Books.
[311] Borchardt, 2011, Einstein's most important philosophical error.
[312] Note that Newton thought light to be a corpuscle as well.

Even when Einstein made odious mistakes in math, indeterminists looked the other way. Steven Bryant[313] and I traced this objectification to Einstein's mathematical somersault in which he first properly derived l (length) from ct (velocity multiplied by time is always distance).[314] Unfortunately, by a little sleight of hand, he then used l, length, as a replacement for t, time. That is how time got to be a dimension. In a way, this has been sanctified in our conventional use of the term "light year," which is the distance light travels in a year. This is an extremely valuable "distance" measurement, but it does not make time a distance.

The General Theory of Relativity reflects this tendency to objectify motion on the grand scale. Time has no dimensions and is not "part" of the universe, although time occurs within the universe. Of course, the concept of "space-time" purports to "combine" space and time. However, only things can be "combined." "Time," having neither three dimensions nor existence, cannot be combined with anything. True, in our heads we can combine concepts, ideas, stories, and equations about real things and real motions, but that does not give them dimensions. A picture of a running dog is not a dog.

"Space" exists, but "space-time" does not. Nonetheless, it is commonplace for cosmogonists to assume that 4-dimensional space-time actually exists. As mentioned previously, this belief is required for the expanding universe hypothesis. Without the mathematical fabrication of space-time, the Big Bang Theory and cosmogony would no longer exist.

---

[313] Brilliant mathematician who forgoed studies toward a Berkeley Ph.D. in physics due to what he recognized early as the regressive nature of the curriculum.

[314] Bryant and Borchardt, 2011, Failure of the relativistic hypercone derivation.

Why has Einstein's objectification of motion been so popular and enduring? I hinted at the reasons for that in my explanation of the deterministic assumption of inseparability. In particular, the belief in motion without matter has been well-established ever since the first humans tried to understand the wind in the willows. That must have been rather frightening. They could not have known that air consisted of unseen particles now called nitrogen and oxygen. Here was a "thing" that was not a thing, like the proverbial ghost that was a thing, but not a thing. The ghost was capable of traveling through walls. It supposedly had three dimensions and location, but certainly was not material. The unseen, unseeable causes of motions were given names—with a god for this and a god for that.

The indeterministic vestiges of the idea of matterless motion are nearly as dominant now as they were in 1905. Matterless motion always has been a mainstay of religion, from holy ghosts, to souls, to gods. Most folks talk about such "things" as if they existed. Thus it was not surprising that Einstein and many others would objectify time to great applause. Modern physics is founded, not on the assumption of inseparability, but on its indeterministic opposite. Einstein and followers never understood this. Make no mistake about it. Time cannot dilate and space-time does not exist. Many of the paradoxes and absurdities in modern physics and cosmogony are traceable to this single most critical philosophical error. We can do better, but only if we give up the idea of matterless motion.

PARADIGM. A Greek word expressing a pattern of thought that generally includes a prominent theory and its associated corollaries used to interpret certain aspects of the world. With respect to science, Thomas Kuhn defined a paradigm as "universally recognized scientific achievements that, for a time, provide model problems and solutions for a community of practitioners."[315] Examples include neo-Darwinian evolution, plate tectonics, relativity, and the Big Bang Theory. Other terms for it include: "the conventional wisdom" or "mainstream thought." Practitioners closely guard the paradigm, usually rejecting non-compliant papers and proposals for employment or financial support that do not conform. Work within a paradigm is considered "normal science," whereas largely unsupported work outside it is usually considered "crackpot," if unsuccessful or "revolutionary," if successful. As with all revolutions, the overthrow of one paradigm by another is considered as difficult as it is monumental.

POSITIVISM. A branch of the philosophy developed from empiricism that assumes that things unsensed do not exist. One variant, operationalism, assumes that things not detected through some mechanical operation do not exist. Thus, positivists reject theism, metaphysics, and speculation that they regard as having no basis in prior experience. They generally assume that space is perfectly empty despite the lack of evidence for such and the theoretical necessity for the existence of aether.

Like classical mechanics and classical determinism, positivism was initially of great value in combating indeterminism. It is still used today when atheists assert that a particular god does not exist because they cannot find evidence for it. Unfortunately, this approach tends to fail with advances in instrumentation. We cannot see the nitrogen, oxygen, argon, and carbon dioxide in air, but now we can detect them just fine with instruments. Today, positivism is found mostly on the indeterministic side of the philosophical struggle. For instance, religious positivists use the "god of the gaps" argument to claim that missing transitional

---

[315] Kuhn, 1996, The structure of scientific revolutions, p. 10.

fossils in parts of the sedimentary record are proof that evolution is false. And then, of course, aether denial is nothing, if not, positivistic.

PROGRESSIVE PHYSICS. The deterministic version of physics that will replace the regressive version that became dominant with the increasing popularity of Einstein's relativity. There always have been regressive elements within physics, but the obvious paradoxes and contradictions brought forth by relativity now bring the regression to a head. Matter-motion terms eventually will be seen as what they are: calculations, not as things or motions. Progressive physicists assume *infinity*, believe that light is a wave, that aether exists, and that momentum, force, energy, and space-time do not. Most do not believe Einstein's Untired Light Theory and that the universe is expanding, and had a beginning via explosion from nothing in violation of *conservation*.

REDSHIFT. The process by which wavelength becomes longer and/or frequency decreases. The "red" in the term implies an increase in wavelength derived from the fact that low-frequency red light has a longer wavelength than high-frequency blue light. It is a bit of a misnomer because color is determined by frequency, not wavelength. There are many different types of redshift as explained in the text. When wave velocity is constant, the relationship: $v = \lambda f$ remains constant, where v = velocity, m/s, λ = wavelength, m, and f = frequency, cycles/s.

REFORMIST PHYSICS. Any attempt to resolve the paradoxes and contradictions of modern, regressive physics by modifying relativity without discarding its indeterministic assumptions and interpretations entirely. Ever since relativity became popular, thousands of mostly unfunded skeptics have voiced objections and proposed modest and sometimes immodest alternatives. Its overthrow is the grand prize among those who view its paradoxes and contradictions as signs that relativity is ripe for the picking. In 2012, Jean de Climont of France developed a list of more than 8,000 dissident scientists with some presence on the Internet who objected to relativity and quantum mechanics.[316] His 2016 version includes over 700 alternate theories, which Climont says are "all amazingly very different."[317] There are over 550 alternate theories that use aether alone. Bet you never heard of any of these. Can you see why the media tends to shy away from any one of them? That is like the legal attack that dumps truckloads of irrelevant documents on the defense. Of course, reporters with even a smidgeon of knowledge about physics and cosmology are rare. The knowledgeable ones need to defend mainstream theories that they have already promoted. In the interest of sales, reporters must confirm the views of their audience. Efforts to destroy those views will not be met with open arms. In any case, reporters do not have the time or interest to sort through hundreds of alternatives to what they firmly believe anyway.

Reform, of course, is not up to the media. In physics and cosmology, the switch from one paradigm to another is the job of physicists and cosmologists. Nonetheless, as in the free will debate, the reform discussions currently are interminable. One wag even summed it up with something akin to the Second Law of Thermodynamics: "Discussions about Special Relativity naturally and quickly degrade into disorder and nonsense." Nonetheless, folks continue to seek compromises that might leave enough of relativity and cosmogony to be acceptable to the mainstream. Above all, one must be able to understand the numerous Einsteinisms in which relativity got the right answers for the wrong reasons. In other cases, his interpretations, such as the Untired Light Theory, are just plain wrong.

---

[316] de Climont, 2012, The Worldwide List of Dissident Scientists.
[317] de Climont, 2016, The Worldwide List of Dissident Scientists.

Except for the dissident press, manuscripts unfavorable to relativity or the Big Bang Theory normally get the circular file. Partly this is because much dissident work is dreadful, not amounting to much more than the kind of silly modifications suggested by funded practitioners. Some of it is overtly religious or entertains other outrageous propositions. I have attended dissident talks proclaiming that the biblical flood covered most of the western US and formed the Grand Canyon. A few still insist that the Sun revolves around the Earth. One prevalent complaint about the dissident community is that the members seldom cite one another and that there is little co-authorship and that fundamental assumptions are rarely stated. Nonetheless, there has been much fine work done by a select few dissidents, Sagnac, for instance. Einstein has been proven wrong as often as he has been "proven right." These instances receive little publicity from the popular, indeterministically flavored press. As always, the main problem with reform is that it does not go far enough. As with agnosticism generally, mixing progressive elements with regressive elements will not remove the contradictions in interpretation.

REGRESSIVE PHYSICS. The indeterministic version of physics otherwise known as "modern physics." I coined this phrase to describe the radical 20th-century departure from determinism that became popular after Einstein's invention of Special Relativity Theory in 1905. The paradigm still has a solid grip on the discipline despite being plagued by numerous paradoxes and contradictions in the interpretation. Practitioners are heavily financed and allowed to publish outlandish deductions and wild speculations as long as they do not contradict relativity and cosmogony. Among these are the concepts of massless particles, immaterial fields, wormholes, multiverses, space-time and other fabrications, such as string theory that claim anywhere between one and 26 dimensions.

Regressive physicists generally do not know what time is and that it cannot dilate. They believe that momentum, force, energy, and space-time actually exist, and that aether does not. Most do not know the proper interpretation of Maxwell's $E=mc^2$ equation. Regressives unwittingly accept Einstein's Untired Light Theory and the resulting interpretation that the universe is expanding and had a beginning via explosion from nothing. Today, almost all gainfully employed physicists appear to be regressive, with dissenters having been weeded out long ago.[318]

The regression began in response to the deterministic ravages of classical mechanism and dialectical materialism during the late 19th century (Darwin, Marx, etc.). In particular, Lenin's "Materialism"[319] and the rise of communism put the fear of god into the west. Relativity's immaterialistic assumptions and paradoxes fit in with the religious beliefs of the day. Einstein prepared the way for the expansionist movement and the good priest Lemaître suggested this meant that the entire universe exploded from a "cosmic egg" in tune with Genesis. To get popular, all ideas, theories, papers, and books must fit the macrocosm in which they exist. Each acts as a weapon in the philosophical struggle between determinism and indeterminism, which continually involves spiralic progress through education and regress through miseducation. When first introduced, the absurdities in Special Relativity Theory and General Relativity Theory brought forth numerous complaints, but these were dismissed in favor of the financial rewards available to their supporters. When viewed with indeterministic eyes, falsifications of relativity were ignored, while so-called confirmations were revered. Opponents usually were dismissed as "cranks" or "crackpots" no matter how logical their arguments may have been.

---

[318] http://go.glennborchardt.com/Borchardt12censorship
[319] Lenin, 1908, Materialism.

What makes relativity so complicated is that some of the predictions have been confirmed even though the interpretations may be incorrect. The challenge for neomechanists is to find the physical causes for those confirmations.

SOLIPSISM. The self-centered belief that the existence of the universe depends on consciousness. Solipsists are immaterialists, who in the extreme might even think that the entire external world depends on them and then disappears when they do. Einstein was being solipsistic when he claimed that fields were "immaterial." Solipsism is the indeterministic opposite of materialism.

SPACE. An abstraction for matter that has less mass than its surroundings. As with all abstractions, there is no such thing as "space," there only are individual examples of space, such as what lies between the electron and the proton, between doorposts, or between galaxies. There can be no "perfectly empty space" or absolute vacuum devoid of matter.

It might be helpful to think of "empty space" as though it were a scaled-down Milky Way. No matter how small the scale, there is always some matter (like the stars) separated by what at first might seem to us as "empty space" (like the interstellar regions). This is the essence of the consuponible assumptions of *infinity* (The universe is infinite, both in the microcosmic and macrocosmic directions) and *interconnection* (All things are interconnected, that is, between any two objects exist other objects that transmit matter and motion). As mentioned many times before, in Infinite Universe Theory we realize that "perfectly solid matter" and "perfectly empty space" are only idealizations. The reality always is something in between. This implies that non-existence is impossible for each part of the universe, no matter how large or small. It is impossible for the Infinite Universe not to exist. The nothingness that indeterminists have imagined is only an idea. There are two things that the Infinite Universe cannot produce: perfectly solid matter and perfectly empty space.

Again, our slogan that "space is matter" is based on the observation that no portion of the universe is completely void of matter. We cannot produce a perfect vacuum and the 2.7°K Cosmic Microwave Background tells us that even intergalactic space contains microcosms in motion. The completely empty space assumed by Einstein and other aether deniers would have a temperature of 0°K. At the other end of the continuum, black holes, if they exist, could not contain "solid matter" without "empty space." They would be subject to Maxwell's $E=mc^2$ equation—they would emit radiation just like all other microcosms. Even Stephen Hawking finally admitted that this is true—they are not black, like he first assumed, but grey.[320]

SPACE-TIME. A matter-motion term for an idealization or visualization of the location of things with respect to the past or future. Note that, like all matter-motion terms, space-time neither exists nor occurs; it is a concept.

SPECIAL RELATIVITY THEORY (SRT). Einstein's unattributed development of Maxwell's 1862 $E=mc^2$ equation in which he erroneously considered light to be a massless particle having constant velocity. His theory of light required eight absurd ad hocs (Table 6). SRT makes plentiful use of the Lorentz factor, γ, which has been used to support solipsistic claims of time dilation, increased mass, and decreased length for objects in motion through empty space. SRT is a mixed bag, with the $E=mc^2$ equation being correct and many of the indeterministic claims being incorrect. For example, time, t, and length, l, are not

---

[320] Lewis, 2014, Grey is the new black hole.

interchangeable categories;[321] energy is not equivalent to mass, time cannot dilate, aether exists and the photon does not, etc.

STEADY STATE THEORY (SST). The reformist theory developed by Bondi, Gold, and Hoyle[322] that attempted to reconcile the expanding universe interpretation and the idea that the universe was infinite and had no beginning. The "steady state" part proposed the creation of enough hydrogen atoms out of nothing to be the driving force for the expansion. Like the Big Bang Theory itself, this was a violation of the First Law of Thermodynamics. Cosmogonists, of course, were not happy with SST even though the Big Bang Theory suffered from the same malady. To this day, indeterminists commonly mistake SST for the true Infinite Universe Theory, which is the subject of the present book.

SUBMICROCOSM. A microcosm that exists inside another. A submicrocosm is always smaller than the microcosm that contains it.

SUPERMICROCOSM. A microcosm that exists outside another. A supermicrocosm can be either larger or smaller than the microcosm. The "super" prefix is used only to designate position.

SYSTEMS PHILOSOPHY. The current scientific world view or philosophy of science that treats portions of the universe as systems more important than their surroundings. The archetype of systems philosophy is the Big Bang Theory.

TELEOLOGY. Any statement that ascribes purpose to objects. This would be a teleological statement: "The boulder wanted to roll down the hillside." Although it still would be correct to state

---

[321] Bryant and Borchardt, 2011, Failure of the relativistic hypercone derivation.
[322] Bondi and Gold, 1948, The steady-state theory of the expanding universe; Hoyle, 1948, A new model for the expanding universe.

that the boulder was compelled to roll due to erosion and gravitation, the word "compel" is most often used when humans are involved, as in "The physics professors compelled her to believe in the Big Bang Theory."

TELEPHONE CHAIN. Game in which a statement is whispered from one person to another in sequence. Due to imperfections in speaking and hearing, the end result of a long chain often can be quite amusing. In science, it has numerous analogies whenever microcosms or their motions are being reproduced. Examples include cell reproduction and the wave reproduction that produces the cosmological redshift.

TIME. Motion. Universal time is the motion of all things with respect to all other things. Specific time is the motion of one thing with respect to another. All events and changes are motions. Clocks measure motion and would be impossible without it.

UNIVIRONMENT. A microcosm and its macrocosm. A word coined by Elizabeth Patelke and myself to overcome the microcosmic overemphasis of systems philosophy (of which the finite universe is the archetype). It was first printed in the 1984 review manuscript of "The Scientific Worldview."[323] The unification also was a rejection of "environmental determinism," with its overemphasis on the macrocosm.

UNIVIRONMENTAL DETERMINISM. The philosophy and universal mechanism of evolution based on the observation that what happens to a portion of the universe is determined by the infinite matter in motion within and without. In addition to being the universal mechanism of evolution, univironmental determinism also was proclaimed to be "The Scientific

---

[323] Borchardt, 1984, The Scientific Worldview.

Worldview" in the review manuscript in 1984[324] and the published book in 2007.[325]

UNTIRED LIGHT THEORY (ULT). The belief that, light, unlike other things or motions can travel from one point to another without losses, in contradiction of the Second Law of Thermodynamics. The opposing belief is called the "Tired Light Theory," which states that the cosmological redshift is due to imperfect reproduction of waves as they travel through the aether medium. ULT is required for the interpretation that the universe is expanding.

VORTEX THEORY (VT). A theory that emphasizes the prominence and importance of rotation in the formation of cosmic bodies. Stellar dust clouds, in particular, tend to aggregate into solar systems only when they begin to rotate after being sideswiped by other clouds. Vortex formation follows Stokes' Law.[326] This law is used constantly in studies of soils and sediments, in which we determine particle size distribution by suspending sediment samples in water and allowing them to settle under the push of gravity. It is simple: gravels fall first, then sands, then silts, and then clays. By measuring the amount of material in suspension as a function of time, we can prepare a particle distribution curve. Vortices demonstrate the effect when their rotation accretes matter by size and density. Over time, the center becomes increasingly dense, with ever smaller and less dense microcosms lingering in the outer reaches.

---

[324] Ibid.
[325] Borchardt, 2007, The Scientific Worldview: Beyond Newton and Einstein.
[326] Puetz and Borchardt, 2011, ibid, Eq. 12.6.4: $vp = (2/9) \cdot gr^2 \cdot (\rho p - \rho m)/\mu m$
Where: vp = upward or downward velocity of a spherical particle
  ρp = particle's density
  ρm = medium's density
    g = gravitational acceleration,
    r = radius of the spherical particle, and
  μm = dynamic viscosity of the medium.

Appendix

Speculative properties of aether

Calculate the mass of an aether particle:

From the neomechanical perspective, I speculate that the "smallest quantity of motion" would be one "cycle" in the Planck equation:

$E = hf$

Where:

$h = 6.62606957 \times 10^{-34}$ m² kg / s

f = frequency, cycles/s

For one cycle, that leaves:

$E = hs = (6.62606957 \times 10^{-34}$ m² kg / s$)(s)$

$= 6.626 \times 10^{-34}$ m² kg

$= 6.626 \times 10^{-34}$ m² kg $\times 10^3$ g/kg

$= 6.626 \times 10^{-31}$ m² g

If we assume that one cycle involves the collision of one aether particle with baryonic matter,[327] then we can calculate the mass of that aether particle from its kinetic energy:

$E = \frac{1}{2} mv^2$

Where:

---

[327] Because light is a wave in the aether, I am assuming here that the collision demonstrating this "smallest unit of motion" is a result of the lead particle in the wave. The collision could not be produced by a photon, because a photon supposedly has no mass.

v = aether particle velocity = $c \times 1.5$

The factor 1.5 is an analog from the ratio of the particle velocity of nitrogen molecules in air to the speed of sound (515 m/s divided by 343 m/s).[328] In other words, the ability of a medium to conduct wave motion is dependent on the velocity of the individual particles that comprise the medium. That also is why a medium can produce "instantaneous," high-velocity wave motion when its motion is triggered at the source. In the case of sound, the individual particles are already traveling at velocities that are 50% greater than wave velocities that occur in air.

Leaving:

$E = (1.5 \times 0.5) \times mc^2 = 0.75 \, m \, c^2$

Rearranging:

Mass of an aether particle:

$m = E/(0.75 \, c^2) = 6.626 \times 10^{-31} \, m^2 \, g \, / \, ((0.75)(3 \times 10^8))$

$= 6.626 \times 10^{-31} \, m^2 \, g \, / \, 6.75 \times 10^{16} \, m^2$

$= 6.626 \times 10^{-31} \, g / \, (6.75 \times 10^{16})$

$= 0.98 \times 10^{-47} \, g$

Note that this value is a maximum because the calculation assumes that only one particle is involved. Because the collision is a wave, many particles, just not the lead particle, might be involved. Then, the mass calculated for each of them would be considerably lower.

Calculate the number of aether particles in an electron:

The known electron mass is 9.109×10^{-28} g and the known classical electron radius is 2.8179×10^{-13} cm.

---

[328] Holtzner, 2016, Using the Kinetic Energy Formula to Predict Air Molecule Speed.

Thus, the volume of the electron would be:

$$V = \left(\frac{4}{3}\right)\pi r^3$$

$V = 1.33 \times 3.1416 \times (2.8179 \times 10^{-13})^3 \text{ cm}^3$

$= 4.189 \times 22.376 \times 10^{-39} \text{ cm}^3$

$= 93.732 \times 10^{-39} = 9.3732 \times 10^{-38} \text{ cm}^3$

The number of aether particles comprising the electron would be:

n = electron mass/aether particle mass

$= (9.109 \times 10^{-28} \text{ g})/(0.98 \times 10^{-47} \text{ g})$

$= 9.295 \times 10^{19}$

Calculate the dimensions of an aether particle:

As we saw, because the aether medium transmits mostly T waves, the particles comprising aether must be vortices (Figure 40). Let us assume that these discs are about 5% as thick (h) as they are wide. Here is the equation for the volume of a disc:

$$V = \pi r^2 h$$

As mentioned, the classical electron radius is 2.8179×10^{-13} cm and the volume of the classical electron is 9.3727×10^{-38} cm^3. Also, as calculated, the number of aether particles in an electron is 9.295×10^{19}.

The volume of a square aether particle would be:

V = volume of an electron, cm^3/number of aether particles in an electron

$= (9.3727 \times 10^{-38} \text{ cm}^3)/(9.295 \times 10^{19}) = 1.008 \times 10^{-57} \text{ cm}^3$

V = volume of the space taken up by a disc:

$= (1.008 \times 10^{-57} \text{ cm}^3) \times$ (area of circle/area of square)

$= (1.008 \times 10^{-57} \text{ cm}^3) \times (\pi r^2/4r^2)$

$= (1.008 \times 10^{-57} \text{ cm}^3) \times (0.785)$

$= (0.792 \times 10^{-57} \text{ cm}^3)$

Dimensions of an aether particle assumed to be a vortex:

Formula for a disc:

$V = \pi r^2 h = 0.792 \times 10^{-57} \text{ cm}^3$

At the assumed ratio of height/radius = 0.1, the equation becomes:

$V = \pi r^2 h = \pi r^3/10$

Rearrange:

$\pi r^3 = 10V$

$r^3 = 10V/\pi$

$\quad = (10 \times 0.792 \times 10^{-57} \text{ cm}^3)/\pi$

$\quad = 7.92 \times 10^{-57} \text{ cm}^3 /\pi$

$\quad = 2.52 \times 10^{-57} \text{ cm}^3$

$r = (2.52 \times 10^{-57} \text{ cm}^3)^{1/3}$

$\quad = 1.361 \times 10^{-19} \text{ cm}$

$d = 2r = 2.722 \times 10^{-19} \text{ cm}$

$h = 0.1r = 0.136 \times 10^{-19} \text{ cm}$

Calculate the density of an aether particle:

Mass = 0.98×10^{-47} g

Volume = $0.792 \times 10^{-57} \text{ cm}^3$

Density = mass/volume = 1.237×10^{10} g/cm3

Calculate the density of an electron:

d = mass/volume = $(9.109 \times 10^{-28}$ g$)/(9.3732 \times 10^{-38}$ cm$^3)$

$= 0.972 \times 10^{10}$ g/cm^3

Calculate the density of a Neutron:

Mass = 1.675×10^{-27} kg = 1.675×10^{-24} g

Radius = 0.8×10^{-15} m = 0.8×10^{-13} cm

Volume = $V = \frac{4}{3}\pi r^3$

$V = 4.189 r^3$

$= 4.189 \times (0.8 \times 10^{-13}$ cm$)^3$

$= 4.189 \times 5.12 \times 10^{-37}$ cm^3

$= 21.448 \times 10^{-37}$ cm^3

$= 2.145 \times 10^{-36}$ cm^3

d = mass/volume = 1.675×10^{-24} g$/2.145 \times 10^{-36}$ cm^3

$= 0.781 \times 10^{12}$ g/cm^3

Density comparison: aether particle 1.237×10^{10} g/cm^3; electron 0.972×10^{10} g/cm^3; neutron 0.781×10^{12} g/cm^3

CPSIA information can be obtained
at www.ICGtesting.com
Printed in the USA
LVHW080322130519
617600LV00012B/221/P